Becoming an Airline Pilot

Becoming an Airline Pilot

Jeff Griffin

TAB BOOKS
Blue Ridge Summit, PA

FIRST EDITION
FIRST PRINTING

Copyright © 1990 by **TAB BOOKS**
TAB BOOKS is a division of McGraw-Hill, Inc.

Library of Congress Cataloging-in-Publication Data

Griffin, Jeff.
    Becoming an airline pilot / by Jeff Griffin.
      p.   cm.
    ISBN 0-8306-8449-2
    1. Air pilots. 2. Aeronautics—Vocational guidance. I. Title.
    TL561.G75 1990
    629.132′5216′023—dc20   90-10910
                                       CIP

TAB BOOKS offers software for sale. For information and a catalog, please contact TAB Software Department, Blue Ridge Summit, PA 17294-0850.

Questions regarding the content of this book should be addressed to:

    Reader Inquiry Branch
    TAB BOOKS
    Blue Ridge Summit, PA 17294-0214

Acquisitions Editor: Jeff Worsinger
Book Editor: Steven H. Mesner
Production: Katherine Brown

Cover photograph courtesy of Boeing Commercial Airplane Group.

Contents

Dedication

THIS BOOK is dedicated to the memory of Joe Christy, one of history's truly great aviation journalists and a man who did me the honor of choosing me as a friend. Joe, formerly managing editor of *Air Progress* magazine and aviation editor for TAB BOOKS, helped me land my first book contract at TAB. Many were the days he would lead me inside his home (kicking his fat little dog Pup-Pup out of the way) and back to his writing den, where we would go over some of the instructional points in my books. Joe enjoyed playing devil's advocate and letting me defend myself against a knowledgeable mentor. Sometimes we would just talk about airplanes and scrutinize old aviation pictures, or discuss whether he should enlarge his home or move to Phoenix.

When our daughter was born, Joe called and surprised us by heaping gifts upon our newborn child. I surely hadn't expected such generosity, and I wound up being more thankful for Joe's true friendship than for the gifts he gave us.

Joe Christy was a consummate aviation writer. If he had wanted, he could have been a consummate aviation novelist; he often spoke of writing a novel. Whenever I see Joe Christy's name on a TAB book, I know that it is quality material written by a most knowledgeable individual whose great love affair with airplanes comes through in every page.

I thank God I knew Joe Christy.

Preface

As a writer it is mandatory that I also read extensively, paying particular attention to books currently available in my various fields of interest. Knowing what's "out there" saves time and helps keep my material current because I won't end up retracing someone else's footsteps. The exception to this is when the subject matter can be improved upon.

This book reflects both approaches: It is in all respects a *new* book, yet it is also an *improvement* on other books that have addressed the subject of preparing for a career as an airline pilot.

The other books I've read do provide some useful information for aspiring airline pilots. They fall short, however, because they highlight the dream and the glory aspects and lightly skim over what the flying life is all about. They paint a picture too much like a Hollywood fantasy.

This book will give you a more accurate feel for what it is really *like* to be an airline pilot. It expands on all the good points you've no doubt heard about, and details the drawbacks as well. For instance, try explaining to a neighbor that you work for a major airline and are based at a city halfway across the country. People just don't get it. Even after you've known them for a while, they still ask you to drop by after work. *After work?* I have a two-hour flight to make after work just to get back to my hometown!

The other books I've seen also fall short as far as specifying a career plan that will get you from where you are now to where you want to be. If you have never been at the controls of an airplane but know that you want to be a professional airline pilot, this book will help you get started. If you already have a pilot's license and have decided to quit the nine-to-five routine for a rewarding career in the airlines, this book will point the way to your goal. No matter where you are or what your situation is, when you finish this book you will have a clear idea of how to take the *next* step and go all the way to the ultimate goal of flying for the airlines.

This book provides a plan, a step-by-step method for getting hired, and information on what you'll encounter on the way to that destination. The price of this book is a small part of what it cost those of us employed by the airlines to reach our chosen goals. I am not saying that this book provides all the answers, but being part of the airline system for a number of years gives me the ability to pass along the knowledge that got me and my colleagues our jobs. I wish I had had a book like this when I was starting out, but I'm glad *you* do; it'll make your career goals much easier to attain.

If you find areas that need improving or updating, please write to me in care of TAB BOOKS and I will consider using your ideas in updated versions of this work. Let me know how you are doing (or have done) on the way to the best job in the world.

And if you end up at Northwest Airlines, look me up.

Introduction

My earliest memory of airplanes and airports is of a Sunday dinner at the Sky Chef Restaurant at an airport near my hometown in Texas. I was five years old, and between bites of the gravy-colored steak my mother had cut up for me, I peered out the window at the intermittent parade of airplanes. As I watched a Trans Texas DC-3 lumber by, engines loping, and the myriad small airplanes (all of which my Dad called ''Piper Cubs'' because nearly all small airplanes back then *were* Piper Cubs) taxiing around, I had no way of knowing what impact this view would have on my life.

That was 1955. In 1956, my interest in airplanes became so keen that I asked my father for a ride on one of those big airliners. He obliged, and at the end of the trip, I was escorted into the cockpit of that glorious DC-3. The cockpit was dark, but filled with an eerie red glow. It was frightening to me and I wanted to turn and run away, but my father held onto my shoulders and started a conversation with the pilots. The pilot in the left seat did all the talking; he was friendly and made it seem okay for me to be there. This was not only my first experience meeting a real pilot, but also my first meeting with someone in a uniform. Those flashy stripes in that dim red light became the next bead on the string of my early airplane memories. In all, it was a very fine birthday present— maybe the finest of my life.

By the early 1960s, I was doing what every other boy was doing— building model airplanes and hanging them from strings attached to the ceiling of my bedroom. The glory of flight was there for me to observe and contemplate in the sanctity of my room. Occasionally, on sunny Sunday afternoons, my father and I would fly gas-powered control-line models in our large front yard. It was ecstasy flying the little models and agony when we had to pick up the pieces that resulted from bad judgment and inexperience.

Those crashes, however, drove me to find out more about *why* the models crashed—and, more importantly, why they *flew*. With great curiosity and a burgeoning love of airplanes, I enrolled in the aviation merit badge program of the local Boy Scout troop. The program was taught by an older Explorer Scout whose father just happened to be the airport manager; they lived in a house supplied by the county right on the airfield. Every Wednesday afternoon during that cool clear autumn, my buddy Bruce and I would ride our bicycles the two miles from school out to the airport for our merit badge class. This became a habit that lingers on today—the almost irresistible urge to visit airports and idly enjoy watching the aeronautical activities.

One day a loud roar echoed through our little Southeast Texas community. "C'mon! It's a jet!" Bruce yelled, jumping onto his bicycle. "Let's go see it!" We raced out to the airport to see what had arrived. It wasn't the first jet in town, but it was the largest we had ever seen.

We arrived at the airport just as the jet was preparing for takeoff. Its two turbine engines whined shrilly as it taxied onto the runway. Then, with an awesome roar, the jet rolled and pointed its nose skyward, leaving us nearly breathless with excitement. The jet age had at last arrived on the wings of a Texas International DC-9-10.

I was in college the next time I found myself standing at the airport fence. Money was scarce and flying lessons were expensive, so I decided I'd wait until I was out of college and had a full-time job so that I could afford to fly. Money was the only obstacle standing between me and my dream of learning to fly, and throughout college, I'd stand at the fence beside the taxiway at the local airport, watching the airplanes and listening to the radio chatter on my aircraft receiver, learning the lingo. Most of all, I hoped desperately that someone would ask if I'd like to take a ride. No one ever did.

This time was not wasted, however. I soon discovered that the college library had every issue of *Flying* magazine from the last ten years on microfilm. Diligently, consistently, I absorbed them all. When the time came to actually learn to fly, this knowledge served me well and expedited my learning process.

After graduating from college with a degree in geology, I soon discovered the workaday world of 12-hour days and the itinerant life of moving from drilling site to drilling site. It was interesting work and I was good at it, rising to an assistant managerial position in short order. Alas, the contrived oil shortage of the mid-1970s shut down oil and gas exploration, and I was out of a job.

With nowhere to go, I took stock of my situation and talents. No one was hiring geologists, so what could I do? I could go back to music and play professionally. I could continue to build my freelance writing career, having already written aviation articles. Or I could take the big plunge and parlay my meager 100 hours of flying time into a flying career.

The decision was easy: I wanted to fly. But paying for the additional flight experience I needed was the hard part.

Recently married and with payments to make on a new house, furniture, cars, and snowmobiles, and *without a job*, my goal seemed impossible. With great determination and help from friends, however, mountains were moved. Nine months later, the ink was fresh on my Flight Instructor license and I began my first flying job, teaching others to fly. Against million-to-one odds, my career had begun.

In this book, you'll learn details about every trick-of-the-trade I had to use to reach my goal of flying for a major airline. Even after my career was airborne, I didn't land that major airline job until I was 35 years old. There are reasons why I didn't do it earlier, and those reasons will be revealed here.

Read this book and carry out this elemental plan of action and *if* you measure up in all the right areas, your rise to the top will be quick.

The airlines are facing a shortage of qualified pilots. If this is your chosen future, then tomorrow will be yours in the airline industry.

1

So You Want to Be an Airline Pilot?

WHAT MADE you pick up this book? Perhaps you want to become an airline pilot—or you are at least interested in what it takes to fly for an airline. The fact that you are reading this shows that you have some interest in becoming an airline pilot, but there is a lot more to a career as a major airline front-seater than you might think. If that is your goal, you're going to need a high degree of interest and motivation in order to make the goal become reality.

The Upside

Before looking at the steps necessary to reach that goal, let's examine what you might expect if you do become an airline pilot.

The average airline pilot is about 40 years old and male. (That's average; opportunities for female pilots today are better than ever.) Within three years of being hired, today's airline pilot will be earning from $45,000 to $75,000 per year at most major carriers. Sound good so far?

In addition to excellent pay and benefits, flying for the ''majors'' has always been an elite profession—not just in terms of prestige, but also earning capability. Let me tell you about a little secret: The average airline pilot earns more than the average lawyer. Remember, I said *average*, and that even includes the dreaded ''B-scale'' or ''two-tier'' pay system, which I'll explain in more detail later.

Apart from the money, there are the airplanes, which are what draw most of us to the airlines in the first place. Sitting in the cockpit and being responsible for the operation of those magnificent flying machines is a dream come true for any pilot (Figs. 1-1 through 1-3).

Also consider the quality of the flight training, which is the best money can buy. Training in the airlines' six-axis motion-based simulators is much more exciting than flying the line—and just as realistic.

Then there is the intangible factor of prestige. You'll notice it the first time you put on your pilot's uniform. As you stroll through the airport terminal, people always watch you. It's not a look of contempt or a stare because you are dressed differently, it's a genuine look of admiration and envy. Take the uniform off and suddenly you're invisible.

Fig. 1-1. American's short 747 (called an SP for "special performance"). American is fast becoming a leader in good employee-management relations as well as already stepping into the number two spot for sheer size of an airline.

Fig. 1-2. Continental Airlines is one of the major employers of new-hire pilots. Whether that affects their safety margin is still being investigated. Currently non-union, look for ALPA to show up on the property any day. When this happens, careers at Continental could prove to be a better deal than they have during the 1980s.

Fig. 1-3. Among the highest-paying and most sought-after employers is Federal Express. Note: They are about to greatly expand their operation. That means hiring.

Fig. 1-3. Among the highest-paying and most sought-after employers is Federal Express. Note: They are about to greatly expand their operation. That means hiring.

Strangers don't say hello anymore (and don't stop you to ask where the bathroom is, either).

Another part of that prestige is the idea that everyone always automatically assumes you have money. For example, you're in a grocery store that you don't frequent often and you're writing a check. The clerk calls the manager to okay the check and the manager notices the checks are imprinted with a picture of one of your employer's airplanes. He looks at you and asks, "You work for Big Wings Airline?"

"Yessir, I'm a pilot for Big Wings," you respond.

He slaps the check back in the clerk's hand and says, "No problem."

It's great! That is, until a new acquaintance asks you to meet him at the country club, assuming you have plenty of money—only he doesn't know you're still on first-year probation.

There's one more important element to being an airline pilot, and that is responsibility. A multimillion-dollar airplane is in your hands when you take the controls. Even if you aren't the Captain and directly responsible to the company for the safe operation of that airplane, you must fly as though you *are* the Captain and responsible for the operation of that airplane.

Many people think that we airline pilots spend hours of sleepless nights worrying about all the passengers riding in the back counting on us. Nothing could be further from the truth. The passengers are safe in the back of the airplane for two reasons: First, we take great pride in flying the airplane as smoothly, safely, and professionally as possible. Second, the crew knows all too well that we are always the first ones *at* the scene of an accident. Read that to mean ''our'' accident.

The Downside

The above factors are the ''up'' side of being an airline pilot. But there are downsides to consider, too.

First, the road to a flying job with a major airline takes a while to navigate. The salaries in the jobs that help you reach the ultimate goal—such as flight instructing, freight flying, charter, and commuter airlines—are not lucrative (Fig. 1-4).

Fig. 1-4. Many pilots cut their flying teeth for regional carriers flying modern, state-of-the-art aircraft.

Another item to consider is the fact that airline pilots must put their livelihoods on the line every six months during proficiency checks and Class I (First Class) physical exams. Flunk the physical and you are out of not only a job but a career—at least until the problem can be corrected, *if possible*. To emphasize this point, consider any other high-paying occupation with this requirement. No doctor or lawyer has to be requalified every six months, but an airline pilot does. Most of us pass our yearly physicals with no trouble, but after the age of 50, the ranks start to thin. Only about 40 percent of all airline pilots make it to mandatory retirement at age 60. Many drop out voluntarily to pursue their own businesses. Some have invested well and become tired of the airline pilot lifestyle. Others just plain flunk out—and I'm not talking about just the physical exam.

There are other considerations. The lifestyle of airline pilot is not attractive to most people. Airline schedules change from month to month and our lines-of-time (pilots' schedules) are hardly ever the same two months in a row. Several factors cause this. Seniority rules and the changes senior pilots make in their bids from month-to-month cause a domino effect. As the senior pilots maneuver to obtain whatever flying schedules fit their needs that particular month, all others below them on the seniority list are affected and must take whatever remains on the schedule. Since deregulation, airlines now change their schedules every six weeks to remain competitive. As flight schedules change, so must the pilots' schedules adapt.

Being away from home (which varies from a small amount to a great deal of time) also must figure into your plans if you are seriously considering an airline profession. This has two sides. For example, many airlines have pass benefits (non-revenue travel), and a pilot is able to live nearly anywhere he chooses. That's a good deal. But at the same time, this calls for that pilot to commute via airliner, and that means more time away from family and friends—*and* the pilot still bears the burden of getting to work on time. This can be scary, but many pilots do it. I do it and problems do crop up, but not as often as you might believe. Commuting is a great benefit, but it can be occasionally stressful.

An important point to consider during these times of deregulated airlines is the two-tier pay scale, or "B-scale." The theory behind the B-scale is that senior employees are entrenched in a certain lifestyle that would be difficult to change without major trauma that would ultimately have a deleterious effect on the airline's operations. To cut payroll costs, rather than penalize the faithful senior employees, a lower starting rate of pay is created for new-hires. The theory here is that career or salary expectations for the new-hires will be lower, and eventually all pilots will be at the B-scale level until they reach a negotiated parity point—say ten years with the company or when upgrading to Captain.

More and more airlines are using B-scales to keep costs down. Here is an explanation of one airline's reasons (actually, the *first* airline to implement B-scales) for using B-scales, given by Robert Crandall, CEO of American Airlines in a speech to the Economic Club of Detroit February 23, 1987:

"Now, in this distinguished company, I'm quite sure I don't have to defend the thesis that people who produce equivalent products cannot bear the consequences of such a huge cost disadvantage for long. Thus, managements at established airlines must bring costs down to something approaching parity with the low-cost carriers.

"At American, we addressed that problem some years ago by creating the industry's first so-called two-tier wage system—an approach which protects the wages and benefits of existing employees and brings new employees aboard at market rates. The alternative is confrontation—the sort of thing that's going on at Eastern these days as Texas Air seeks to slash—by 30 percent plus—the salaries and benefits of Eastern's employees.

"Now, obviously we could take the same approach—and if we must, we will. But we believe that there are considerations other than numbers involved here. An airline is a service business. We think its quality—in every aspect of its operation—depends on the respect with which employees are treated. We just don't buy the theory that you can run a first-class operation by penalizing veteran employees.

"So our game plan is to invest heavily in new aircraft and facilities, using them as tools to expand our company . . . and, thus, to gradually reduce our average labor costs as we increase the percentage of our operation that benefits from the cost advantages of market rates.

"If you accept what I said a moment ago—that any successful company must have at least approximate cost parity with its competitors—then a two-tier system seems to us the most humanistic approach to achieving that goal."

The B-scale is, however, causing problems. It causes animosity in the cockpit, which is not conducive to safety, and makes it difficult to attract the best applicants. Another aspect that is beginning to become apparent is that during expansion of an airline, or when a large number of senior pilots retire, the B-scalers begin to outnumber the A-scalers. Thus the high standard of living the A-scalers sought to protect could eventually be compromised if the airline chooses to cut the salaries of the dwindling number of A-scalers. It's not a happy situation.

Knowing something about the B-scale is important for those considering an airline career for two reasons. Every prospective pilot should know what he or she is getting into. It also helps a pilot in choosing the airline with the *best* B-scale or an airline that *does not have one*. (At this writing, all airlines except Northwest have one, and even the merged Republic pilots at Northwest have a B-scale.) Keep in mind that these schemes change, but knowing about them is an important part of having a realistic attitude about a career as an airline pilot.

Seniority

Most people have no doubt heard of seniority systems. To put it simply, the person hired first is senior to anyone hired after him, and this affords the senior pilot certain benefits. These include more pay, bigger and better airplanes, more days off (if desired), more vacation, more stripes on the shoulder, and better crew meals. In a nutshell, that's the seniority game (Fig. 1-5).

Seniority in the form of a seniority number is assigned on the date of hire (normally). The new-hire class is ranked according to members' birthdays. Therefore, the oldest in the class is most senior and so on down the ranks of those enrolled in the class.

Now, where does the newly hired pilot start on the airline's seniority list? At the bottom of the list, of course—right where most of us *don't* want to be. And this means junior pilots get *less* pay, *smaller* airplanes, *fewer* days off, *less* vacation, and *fewer* stripes. Junior pilots also work holidays and *always* must tell the Captain that they will "gladly" take the chicken dinner (since most airlines require pilots to eat different meals in case of food poisoning).

Fig. 1-5. Seniority is what it takes to get into the left seat of the really big birds, such as this Northwest Airlines 747.

As I mention some of the disadvantages about flying for the airlines, don't get the impression that it's all bad. I'm trying to acquaint you with point number one of this book, which is *building a realistic attitude*. If you are a potential airline pilot and, after reading this book and assimilating this information, can make a good decision, knowing all the facts, then you'll most likely end up happy with your original decision to follow an airline career. Believe it or not, pilots have gone to work for the majors and found it was not what they thought it would be and promptly went into some other business. If your enthusiasm isn't quenched and you want to know more, read on.

Normally, when a pilot is hired by a major airline, there is a period during which the new-hire will have to "sit reserve." The term *reserve* in airline parlance mean "standby" or "on call." A pilot on reserve bids for a schedule just as any other pilot bids for a monthly schedule. Sometimes a reserve block, or line, as they are known, can have advantages because as a pilot's seniority number becomes more advantageous, certain days off can be obtained. A pilot might choose weekends off, for instance. This type of schedule usually would entail four duty days where the pilot is liable to the company to be on call for work, followed by an average of three days off. This varies from airline to airline, but most use a similar method.

The hard part of sitting reserve is having to wait by the phone for the crew schedulers to call with a trip. Most pilots don't like this unless they have projects (such as writing this book!) that can be done at home.

Many reserve pilots carry beepers and move all over town during duty days attending to normal business. Reserve, then, can be a blessing or a curse, depending on your outlook.

Hiring for the Future

Let's take a look at the future and what it holds for pilot hiring projections. Notice I didn't say how hiring *will* be. The first thing you should know about pilot hiring is that it is cyclical. When the economy burps, pilot hiring might be curtailed or even shut down. Even with mandatory retirements increasing through about 1995, we can expect that a major economic downturn will stem the flow of hiring at the major airlines.

In fact, furloughs or layoffs are *always* possible. Historically, however, most airlines have never cut more than 10 percent of their pilot ranks during any economic recession. That illustrates that another point in choosing an airline is to find one that will put 10 percent more new-hires under your seniority number as soon as possible. That's easier said than done!

If you regularly read aviation trade publications such as *Aviation Week and Space Technology*, you will have learned that most airlines have new airplanes on order. Many of these airplanes are going to the freight carriers, which might be a career path to consider. A more important consideration to realize, however, is that the current fleet of Boeing 727s (Fig. 1-6) is aging rapidly and is already starting to be phased out. During this phase-out, not all the 727s will be replaced on a one-to-one basis, because some of the new airplanes serve to expand the airline buying them, or to increase that airline's market share. There should therefore be a net *gain* in pilots needed, even though the new-generation airplanes require only two crewmembers as opposed to the 727's three (Figs. 1-7 through 1-9).

Fig. 1-6. The 727 has been the mainstay of all the airlines for many years. Now its days are numbered as the 727s begin to find their way to the ranks of cargo planes for the indefinite future.

Fig. 1-7. A Northwest B-757 cruises high above the California desert. The 757 is one of the new generation of two-crewmember airliners.

Fig. 1-8. The cockpit of the 757 boasts the latest in technology. It is called EFIS, or Electronic Flight Information System. As you can see, all but a few instruments are depicted on CRTs or televisions.

Fig. 1-9. A closer look at the CRT or EFIS cockpit of the 757. Incidentally, Boeing uses the same cockpit in both the 757 and 767.

An example of a typical airline is illustrated in Table 1-1, which shows the fleet size and projected fleet growth of Northwest Airlines (my employer via the Republic merger). Table 1-1 represents the latest information on Northwest Airlines, as of this writing. Airline fleets are fluid, with airplanes coming and going constantly. According to Steve Rothmeir, Northwest's CEO, "Northwest has historically bought or sold an airplane, on the average, every ten days." With a company history that goes back 60 years, that's a lot of airplanes! One example is that Northwest planned to buy or lease 16 more generic DC-9s and has added a replacement McDonnell Douglas MD-80 for the one lost in a crash in 1987. Ten of the 16 DC-9s were in Northwest service by the end of 1988, with the rest following during 1989. Rumor has it that Northwest is looking at an additional 30 to 45 DC-9s to be obtained overseas (Figs. 1-10, 1-11). And Northwest took delivery of the first of 25 Airbus A320s (Fig. 1-12) it ordered in June 1989. If more Boeing 757s were available, Northwest would purchase 15 of those (Fig. 1-13). Competing airlines, however, have sewn up future Boeing deliveries for several years, and in any case, Northwest is having difficulty hiring and training enough pilots to meet the demand for additional aircraft. Still, that's good news for you pilots out there in the hiring pool.

One thing to keep in mind is that although the chart indicates Northwest will increase its fleet size, some of the company's older airplanes may be phased out. Northwest's 34 DC-9-10s are 20 years old on

Table 1-1. Northwest Airlines Fleet as of September 1988.

	On Hand	On Order
747-400	—	10
747-100/200	32	—
747 Freighter	6	2
DC-10	20	—
A340	—	20[1]
757	27	3
A320	—	100[2]
727-200	71	—
727-100	9	—
MD-80	8	—
DC-9	126	—
Convair 580	13[3]	—
Total	312	135

[1]Not yet confirmed
[2]25 now confirmed additional confirmations through 1993.
[3]Convairs are scheduled to leave for Mesaba Airlines Dec. 1988.

Fig. 1-10. Our new Northwest Airlines livery for the old Republic DC-9s. This one's a Dash 30.

average, its 64 DC-9-30s average 18 years old, and the 28 DC-9-50s average nine years. Northwest's 727-100s are in the 18- to 20-year-old range. Some of the orders for new airplanes will be used to replace older jets mentioned above.

Fig. 1-11. Cockpit of a DC-9-30. Some airlines start their new-hires right here.

Fig. 1-12. An Airbus A300. This one belongs to American.

Fig. 1-13. High over the Montana Rockies an NWA 757 carries 184 passengers and two pilots—one of whom is only in his second year with the airline. Could this be for you?

While Northwest might be phasing out some of its older jets, the DC-9 series (much like its older sister, the renowned DC-3) is a mainstay of the domestic short-haul system and very capable of serving well into the 1990s. The Boeing 727, on the other hand, is rapidly becoming obsolete because of the required extra crewmember—the Flight Engineer.

Larger airlines, such as Northwest or American Airlines, which generally pay higher wages than, say, Continental or America West, must cut costs in any way possible, and that means getting rid of the airplanes that are the most expensive to operate.

To illustrate that point, Northwest is selling its 727s even though the DC-9s are just as old. The reason is the difference in number of engines and required crewmembers. American Airlines and others are adopting a similar strategy. American, for instance, is phasing out its 727-100s and bringing on masses of MD-80s (Fig. 1-14). Sources at American claim a pilot force of 7,000 and that the airline is still hiring. Having purchased 100 airplanes in 1987 and another 66 in 1988, the end of pilot hiring is nowhere in sight.

How do additional airplanes affect hiring decisions? Take another look at Table 1-1. Every airline has its own staffing requirements for its airplanes. The number of crews required for each aircraft is contractual in nature, so some airlines might require more crews than others. The numbers generally fall between three and six crews per aircraft, with five being near normal (but not necessarily average). Applying those numbers to the table, we can see that Northwest has 135 airplanes on order. At five crews per airplane and two pilots per crew, this would amount to 1,350 pilots for the new airplanes.

Unfortunately, we also know that the 727s (80 total) are leaving and 16 more DC-9s will be arriving. The net increase is more like 71 airplanes, making the total pilot needs more in the area of 800 to 900 when you factor in the Flight Engineers that are already hired and must be absorbed in the system. On the other hand, we do know that retirements will be increasing, and that again pushes the number of needed pilots upward.

Fig. 1-14. During the mid to late 1980s, new-hires were moving into the right seat of MD-80s such as this one within a year of their hire date. American's on the move.

These numbers basically work the same for all airlines. There isn't room in this book to include all airlines and their future plans, and in any case their strategies change constantly depending on the state of the economy. As an up-and-coming airline pilot, you can stay informed by consulting *Aviation Week and Space Technology*. The subscription price might be a bit high for a ground-floor pilot, but most large libraries subscribe to this weekly magazine. Another excellent source of airline industry information is *Air Transport World*, although its subscription price is high, too.

All the evidence shows that airline fleets are growing, but why? Are they greedy? Do they just want to fly everywhere so they can compete? The answer is not that simple. If you have been following the airline industry, you have probably heard of Frank Lorenzo, who is largely responsible for the shape the airline industry is in today. In 1981, after taking over Continental Airlines, Lorenzo was successful in filing bankruptcy for Continental, which allowed him to legally abrogate the union contracts then in effect at the airline. This enabled Lorenzo to dictate new pay scales at Continental with no regard to previously negotiated pay rates. Most other airline industry CEOs have denounced this unsavory way of lowering labor costs. It should be noted that the lower pay rates applied to everyone in the airline *with the exception of Lorenzo and his hierarchy*.

The result of all this questionable maneuvering is that Continental ended up with the lowest pay scales and labor costs in the industry, putting the airline in the driver's seat when it came to setting fares. And lower the fares Lorenzo did, causing massive fare wars to break out. Because the other major carriers were not able to reduce their labor costs by dumping labor contracts, they were at a disadvantage in trying to compete with Continental. Something had to be done or Frank Lorenzo would end up with all the marbles.

Airlines solved these problems in two ways. Labor costs were lowered by implementing B-scales for new-hires and cutting wages across the board (in agreement with applicable unions), and route systems were expanded. By flying more miles and (theoretically) increasing worker productivity, labor costs are averaged over a larger route system.

Expansion, therefore, became the battle plan that American's Robert Crandall used to combat Frank Lorenzo and Continental Airlines, and the result of that expansion has been increasing levels of pilot hiring.

Framing Your Expectations

By now you have some insight into the general hiring picture at the major airlines. We'll look at specific figures in Chapter 6, but for now you can see that prospects for future airline pilots look excellent.

The next item you need to consider is yourself and the decisions you need to make regarding your potential career as an airline pilot. First, let's examine what you expect from an airline career and what you will actually achieve should you be hired by an airline.

Before we discuss the salary you might receive, the distinction between *major* and *national* airlines needs to be clarified so you know what you are looking at when you consider applying to specific airlines. What is a *major* airline? After the many mergers of the past decade, a good definition is ''an airline having a vast domestic route system and a well-developed international system with departures from one or both coasts and spanning either the Pacific or Atlantic oceans or both.'' In the U.S., major airlines are:

- American Airlines
- Continental Airlines
- Delta Airlines
- Eastern Airlines
- Northwest Airlines
- TWA
- Pan Am

Large national airlines also pay well. In the U.S., these include:

- Alaska Airlines
- Braniff Airlines (bankrupt again as of this writing)
- America West Airlines
- Midway Airlines
- Piedmont Airlines
- Southwest Airlines
- U.S. Air

Other airlines may fall in the category of ''national'' airlines, but do not have as much national impact as those listed above. The listed airlines are definitely worthy companies at which to seek employment, and regardless of their size, most pay well—and competitively. They *must*, because all pilots are drawn from the same pool.

Just a quick note about the Future Aviation Professionals of America: This organization (which I'll discuss more later) does a great service in providing data to aspiring aviation professionals, especially potential airline pilots looking for jobs. FAPA—for a fee, of course—provides detailed information on which airlines are hiring, what their requirements are, and—most importantly—their pay scales, all of which should help you decide where to apply.

While I'm on the subject of pay in regard to selecting an airline, I want to mention the importance of checking the parity point for B-scalers. Nearly all airlines maintain a two-tier wage system at present. Some, for example, never reach parity until the pilot upgrades to Captain. Some reach parity after ten years, and others do so in much less time. It's important to consider the parity point because it will directly affect your lifetime earning capability.

Salary is not the only thing to be considered in choosing an airline. Another important consideration is *pilot domiciles*. Domiciles are important because not everyone wants to live in the same geographic area. If you want to live in the west, then concentrate on airlines that have pilot domiciles in the west. The same is true for wherever you want to live.

Other factors that rank at the top of considerations for choosing an airline are *retirement programs* and *insurance benefits*. Pilots are always looking for ways to protect their futures.

Once you have chosen a specific airline or two that look interesting, visit an airport and try to speak with pilots who fly for those airlines. Most pilots love to talk about their jobs—even if only to complain! But a few pointed questions to the right pilot can bring many new considerations to light. With this information, you might reevaluate your choice or map out a personal plan of action that will supplement the shortcomings of your target airline. The importance of obtaining this information is twofold. It helps you decide if you are indeed applying to the right airline, and it will show an interviewer that you know something about the inner workings of the company. Both are pluses that will help you land the job.

Subjects such as crew scheduling practices might be too complex to discuss with a friendly pilot and might be tough to track down by simply asking questions of pilots you meet. The expectations, however, should be about the same for each airline. For instance, the news media in recent years has made a big deal about how little airline pilots work. The truth is, we work a great deal, but we get our work done in a small number of days. The news media loves to tout the fact that we pilots work only 75 to 80 hours a month. That *sounds* like just two 40-hour work weeks to anyone who doesn't have a grasp of what really goes on. (No wonder the public is so envious of our positions!) The truth is, however, that it takes 180 to 240 hours a month of *duty time* to fly 75 hours. That's the catch.

Once again, I must emphasize that much of our time is spent away from home. To me, it's a tradeoff: better-than-average pay for abnormal working conditions. We don't work less than the average person; we often work *more*, but only with seniority (*lots* of seniority) does the amount of time spent at work start to look like a better deal than an ordinary job.

Passing the Physical

In qualifying yourself for an airline career, passing the physical examination and getting the required First Class physical certificate is probably the most crucial step. Before pursuing any serious plans for becoming an airline pilot, you should consider this area.

The good news is that wearing glasses or having had surgery will not rule you out of passing the necessary First Class physical exam. The FAA's stance on pilot physicals is that as long as you can perform with an ability equal to others in the profession and your particular ailment does not put the public at risk, then a medical certificate can be obtained.

Here are some firsthand experiences. I have a friend who wears very thick glasses. His uncorrected vision is 20/200. With his glasses, however, he has 20/20 vision. His license stipulates that he must wear glasses to operate an airplane, much like the requirement listed on driver's licenses for people who wear glasses. Now take me, for example. I wore glasses until I was 17 years old. A progressive ophthalmologist (especially for the 1950s) put me on a program of continuously weaker prescriptions, making my eyes stronger and stronger. Eventually I passed the First Class eye exam with no restrictions on my license. The main point here is that wearing glasses need not stop you from an airline career. While it used to be that airlines would only hire pilots with perfect eyesight, the ranks of qualified pilots are now so thin that the airlines are relaxing requirements that used to specify no glasses. Be sure to check your prospective airline's requirements if you need to wear glasses to pass the physical.

As far as physical exams are concerned, no license to fly is valid without a corresponding valid physical exam certificate. The exam can be done only by FAA-designated physicians (Aircraft Medical Examiners). Here are the different classes of medical exams given by FAA-designated AMEs:

First Class: Required for an Airline Transport Pilot Certificate. Valid for six months, but reverts to Second Class for another six months, which is still good enough for most copilot positions at many carriers.

Second Class: Required for Commercial Pilots and Flight Engineer Certificates. Valid for one year, then reverts for one more year to Third Class status.

Third Class: Required for a Student Pilot or Private Pilot Certificate. Valid for two years. Students must possess this exam certificate before first solo.

About the only condition that rules a person out of flying commercially is heart trouble. The FAA is very sensitive on this matter. Even so, many pilots have successfully obtained another FAA-sanctioned physical after heart ailments that have been corrected, although pursuing such a course often results in great expense and personal sacrifice.

Another personal story of mine might encourage you further at this point. Many people would have you believe that any catastrophic illness will rule you out of an airline flying job. Some diseases will, such as nervous disorders, but others, such as cancer, don't always mean instant grounding. How do I know? I developed a testicular tumor during my first year at Republic Airlines. This cancer is highly curable, and thank God I made it through. Today I am still flying and need only submit my oncologist's report with each new physical exam—no big deal. In some ways I am more highly monitored physically than others. That protects me, because I know more about what's going on inside my body, and it also serves to protect the public flying on my airplanes. (Although, I might add, I'm normally harmless to the public anyway!)

The point is, get a First Class physical exam early in your flying career. You don't have to get a First Class physical when all you need is a Third Class to solo, but you should ask the doctor to examine you according to First Class guidelines and let you know how your health stacks up.

Below are listed requirements you need to meet to ensure you are airline material. Not all the listed requirements come from the FAA. The weight requirement, for example, is from a standard doctor's table for determining ideal weight in relation to height. Many airlines use this standard table for qualifying candidates. Exact FAA requirements can be found in Federal Aviation Regulations Part 67.13.

Vision: Corrected to 20/20. With the current pilot shortage, airlines have begun to waive the stiff vision requirements they once cherished so much.

Hearing: Hearing loss in any of the spoken frequencies may not exceed 20 decibels.

Blood Pressure: Average blood pressure must be below 140/90.

Height: Height should be in proportion to weight. For airlines, height should be from five feet six inches to six feet four inches.

Weight: Weight in indoor clothing without shoes should correspond to that listed in Table 1-2.

Table 1-2. Standard Physician's Height-Weight Table.

Height	Men	Women
5' 6"	157	146
5' 7"	161	150
5' 8"	165	154
5' 9"	170	159
5'10"	174	164
5'11"	179	169
6' 0"	183	173
6' 1"	188	178
6' 2"	193	183
6' 3"	199	189
6' 4"	207	197

This information was based on average build
and does not take into consideration frame or body build.

One last item I'd like to pass along about passing the physical exam was told to me by the first doctor I ever saw for the aviation medical exam. As I was filling out the required paperwork before the exam, I came to the part that lists all the various afflictions such as dizziness, diabetes, asthma, headaches, etc. The doctor explained to me that if conditions are not relatively current and chronic, I should not report them. For instance, he explained, when you come to the item "headaches," you must realize that nearly everyone has a headache now and then. Unless you have a history of migraines, the doctor recommends that the box be checked "no." This idea makes sense. Start your career with a clean record and don't volunteer things that might seem innocent to you, but that the FAA could interpret as having a deeper meaning.

Remember, take a First Class medical exam early in your career planning stages. When you begin to fly, you will be required to obtain a Third Class physical before you will be allowed to solo. This certificate must be carried with you and, when endorsed on the back by your instructor, becomes your bona fide Student Pilot's Certificate. Ask the doctor to give you a physical according to First Class guidelines even though you will leave the office with a Third Class in your hand. Tell him your career intentions and he will oblige you. When he gives you the thumbs-up as you walk out the door, you will have surpassed one of the most formidable obstacles to becoming an airline pilot. If you are not able to pass the First Class physical, there is no use spending a dime or a minute of time to accomplish the corresponding career goals.

If for whatever reason you don't pass the First Class exam, you may still be able to fly professionally. A Second Class medical is still adequate to hold down many flying jobs, such as corporate pilot. Although it is hard to envision being able to pass a Second Class physical and not a First Class, there is a distinction between the two and therefore it is technically possible. If you have your heart set on flying for a living, a Second Class can be good enough in many cases. Remember that corporate jobs can be very lucrative and can be even more stable than airline jobs (*sometimes*). Corporate aviation can be a good alternative to airline flying.

Women As Airline Pilots

The common idea of women being the lesser gender in terms of strength has been perpetuated in aviation for a long time. Having mostly women as cabin attendants for so many years has not helped, either. The fact is that some women cannot cut it in the cockpit because they are not strong enough to handle an aircraft that has lost all power augmentation to the flight controls and has reverted to manual-only operation.

There are two types of women in airline flying jobs: those who endeavor to become successful and those who don't. Any woman interested in flying for an airline will find it necessary to overcompensate, to try harder to make it in a field dominated by men. A woman considering a predominately masculine occupation must know that she's in the minority.

The question is, what will that mean to her overall endeavor in aviation? It means with good qualifications, she will be hired. Airlines have hired many women since the 1970s, and some now occupy left seats. With few women applying for the job, and equal-opportunity employment laws in force, many companies are attempting to hire a certain percentage of female pilots. If you are a woman, your chance of being hired is much greater than any male applicant. The ladder to gain the required qualifications, however, is the same ladder that must be climbed by your male counterparts. It takes patience, ability, desire, and money to get there. Thus anything written here applies to both female and male airline aspirees.

Time and Money

Can you put the money and time into learning to fly? This is the second hardest question in preparing yourself for a possible flying career with the airlines. *Flying is not cheap!* The cost of flying can be minimized if you work at it diligently and look for ways to do just that. Regardless, there are, as you might imagine, minimum FAA requirements that must be met.

Let's look at what flying costs in the civilian market. Rates vary from area to area and usually never go down. At this writing, I live in Colorado Springs (Colorado), and I checked the rates at a local flight school. (I suspect that rates in Colorado Springs are far from the least expensive, yet not at the high end compared with some major metropolitan areas. One thing to keep in mind, however, is that rates are usually very low in areas of good weather. Los Angeles is a great example of this phenomenon. Rental rates are extremely low in the Los Angeles area because there is such an incredible amount of competition. So be sure to shop around as much as possible.) I found the following nonclub rates in effect at the Colorado Springs flight school:

- Cessna 172 Skyhawk (four seats, fixed gear): $51.00 per hour
- Beech Sundowner (four seats, fixed gear): $51.00 per hour
- Piper Arrow (four seats, retractable gear): $65.00 per hour
- Flight Instructor: $18.00 per hour

When learning to fly, a great deal of time is spent with a flight instructor. Thus we must add the $51 per hour figure to the $18 per hour instructor figure for a total of $69 per hour for dual (with an instructor) flight instruction. I think this is steep!

It quickly becomes apparent that certain short cuts or different approaches must be found. One way is to fly smaller airplanes, such as

Cessna 152s or Piper Tomahawks (two-seaters), which rent for up to 25 percent less than the above airplanes.

Let me put this into a little clearer perspective. Let's assume a Cessna 152 rents for $38 per hour (an actual current rate in many areas of the country, including California) and add the above rate for flight instruction of $18 per hour for a total of $56 per hour. Part 61 of the Federal Aviation Regulations requires a student pilot to fly *at least* 40 hours to qualify for a Private Pilot's Certificate. Of those hours, 20 must be dual (with an instructor). So at current rates, the airplane alone will cost $1,520. The instructor will cost an additional $360 just for the flying time. The total comes to $1,880. Please note that this is a *minimum*. Some students can complete a course in the minimum time, but most cannot. In any case, there are added costs for ground school, flight planning equipment, etc.

Another option is to attend a so-called certified flight school. These operate under a different part of the Federal Aviation Regulations (FARs) called Part 141. The FAA allows these schools to lower the minimum hours required for a Private Pilot's Certificate to 35 flight hours total. This does not allow reduced instruction time, but does allow less solo time (when the student is the sole occupant of the airplane). In financial terms, this would lower the overall cost to $1,690, a savings of $190—that is, if the Part 141 school offers the same rates as the Part 61 flight school, which may not be the case. Also, the Part 141 school might require that you attend their ground school, so there goes the money you may have saved.

The big difference between Part 61 and Part 141 schools is that Part 61 schools usually serve a clientele who wants to learn to fly at their own pace, once or twice a week after work or school and on weekends. If you have to work while learning to fly (in order to pay for it), then this may be the way to go.

Part 141 schools, on the other hand—especially the large organized companies that advertise in all the aviation magazines—specialize in training students, many of them from foreign countries, rapidly and in a minimum amount of calender time. Students at these schools frequently bury themselves in flying, almost every day, in order to obtain their Commercial, Instrument, Multiengine, and Flight Instructor Certificates in as short a period of time as possible. Because these schools cater to such a committed clientele (many of whom are sent by foreign airlines and governments), their rates for airplane rental frequently are at the higher end of the scale.

There are many different types of schools that fall between the two descriptions above, and the only way to find the right one for *you* is to shop around based on *your* special needs. I'll mention some other options you might want to consider in the next chapter, but first you need to know the basic minimum levels of qualification you must obtain to get started in this business.

Beginning as a Private Pilot is just that—the beginning. If you are to be worth anything to an employer, you must acquire a Commercial Certificate with Instrument rating. You can obtain these within 200 hours total time under Part 61 and 160 hours under Part 141. To build that time by paying for it hour by hour can be prohibitive. Even at the lowest rates, about $5,500 is needed to reach a level that might *or might not* be high enough to find a flying job.

There are several ways to tackle the problem of raising the money. These involve lending institutions, financing through the flight school, college grants/student loans, the time-honored working in exchange for flight time and, of course, paying with cash (if you have it). We'll examine some of these and other methods in the next chapter, but at least now you know basically what you're up against financially when learning to fly.

Now you have to consider this question: Can you devote the *time* to earning all those licenses? Most people can if money isn't a problem. What the question *really* implies is, will you as a candidate for an airline piloting career do what is necessary to stay current and sharp as pilot? Fly only the minimum required three takeoffs and landings every 90 days and you'll assure yourself of never becoming much of a pilot, professional or otherwise.

Everyone is different, but flying a minimum of once a week during the Private Pilot stage is necessary. For someone who is truly motivated and eager to learn, twice a week is adequate and just about right. From my experience as a Flight Instructor, I'd say that more than this amount will result in no better retention and progress, unless you can afford to devote full-time to learning to fly. After you have your Private Pilot's Certificate, the amount of flying can be increased. Even so, at these modest rates, the average person will take one to three years to complete the flight training curriculum needed to become *minimally* employable as a commercial pilot—which offers very few opportunities for low-time pilots except for jobs such as glider towing, banner towing, pipeline patrol, fish spotting, etc. To become more desirable as a working pilot, it will be necessary for you to obtain a Flight Instructor rating and Multiengine rating. Both, of course, require more money and more time.

Civilian or Military Route?

This question invariably depends on the makeup of the individual. Some people can join the military, endure all the military hoopla (separation from loved ones, authoritarianism) and come out feeling like they had a great experience and fun to boot. Some cannot. This one point aside, there are points that may tip the balance one way or the other for you.

The question always returns to time and money, but that's the way most desirable things are in life. Consider this: A military career begins in college and even a civilian's flight career should entail a college degree because most airlines demand that. That is changing somewhat, but in any case, military and civilian routes are equal up to that point.

Once a person opts for a military career, however, getting to the airline job will take the length of his commitment to the service of his choice, which is a minimum of six years. This cannot be shortened. The civilian route entails the penalty of finances for the freedom, but the time can also be cut much shorter with proper planning. For example, your flight training can run concurrently with college. In fact, you could complete your training for your Flight Instructor rating before completing college and start teaching others to fly and building flight time before graduating.

In fact, you can attend aviation universities that grant aviation-related degrees and learn to fly as part of the curriculum. These schools usually hire students who have earned their Flight Instructor ratings to teach new students to fly, thus giving instructors the opportunity not only to build flight time but also to earn some money to help defray the high cost of their education.

You might, however, want to think about the ''all-your-eggs-in-one-basket'' theory. Depending on how you look at life, it might be more important for you to major in another area besides aviation in order to have a backup if flying becomes no longer possible. It's up to you which way you go. Nonetheless, the flight time a civilian pilot can build prior to and during college might be sufficient to put him in the airline hiring pool as early as two years out of college—if not even sooner, for an over-achiever.

On the other side of the coin, a military career looks attractive. After all, you receive excellent training. Military training syllabuses are consistent and change only with new aircraft or airborne equipment, resulting in a steady, high-quality program that results in excellent pilots. Smaller civilian flight schools tend to be less consistent because personnel are constantly turning over. Civilian training at the small flight school level is more a function of individual instructor technique than tried-and-true curricula. At the larger aviation universities, curricula and techniques are somewhat more standardized and closer in training quality to their military counterparts.

Some of the fine civilian schools with excellent reputations include Embry-Riddle, FlightSafety, American Flyers, Daniel Webster, and Sierra Academy, just to name a few. There are many others, and if you look through any current aviation magazine, you'll see their advertisements.

As I often do, I have wandered away from the subject, which is the attractiveness of the military. Other than quality of training itself, there is the variety of aircraft—jet fighters, heavy bombers, and troop and cargo transports. All would be exciting choices. From what I've seen, most military backgrounds of airline pilots tend to be in fighters. It would seem that a transport job would better prepare a pilot for an airline career than a fighter assignment. I guess a lot can be said for sowing those wild oats and appeasing the "need for speed," as Tom Cruise said in the movie *Top Gun*, while young and more able to perform that sort of service for one's country. After all, you can still get an airline job after flying fighters and fly those *slow* transports later.

The main drawbacks to choosing the military direction are two. Obviously, there is the amount of time that you must enlist. Six years can be a *long* time and if the airlines are in a hiring mode, then many hundreds if not thousands of seniority numbers will be missed. Don't underestimate this fact. As I discussed earlier, *seniority is everything*. The second drawback is that military pay, including benefits, is not too bad. It's not that the airlines cannot beat the military rates of pay; they will. However, coming to the airlines from the military, there's that probationary year and those lower pay rates until you are well entrenched in the airline system. In other words, a financial adjustment will have to be made when separating from the service. It's something to think about and prepare to face (Fig. 1-15).

American Airlines

Fig. 1-15. There was a time when most all pilots came from the military before they landed a job in the cockpit. That no longer holds true. Cockpit here is a 747 SP.

Another Alternative

In the 1960s, pilots became a scarce commodity. You might have heard stories of Private Pilots stepping out of their Cessnas and coming face-to-face with an airline representative. As the story goes, those low-time pilots with no previous experience or designs of becoming an airline pilot were offered training and jobs at the major airlines' expense. Imagine the airlines being so desperate that they would pick up the tab for training and offer a plush, soft-seated job in the front of a large jet! I assure you, that *did* happen—and we might see that again in the 1990s. Two airlines—United and Northwest—are thinking that this might become a reality and are posturing themselves to meet the demand for pilots. United already has a large training center in Denver, Colorado, which is used to train United pilots and pilots for many other airlines. It could be used to train neophytes, and United is getting ready for it. Northwest, on the other hand, has taken an even bolder step.

Northwest Airlines announced in 1987 that it had made a $70-million commitment to buy more than 21 flight simulators. Facilities were built that could be enlarged to handle up to 32 full-motion flight simulators. Entwined in this plan is the University of North Dakota, which has agreed to develop training programs for pilot candidates from zero-time to Airline Transport-licensed. This agreement will make it possible for the university's Center for Aerospace Sciences to meet Northwest's pilot needs as well as a burgeoning need for airline pilots around the world.

The university, at Grand Forks, North Dakota, already operates 70 aircraft and has more than 70 students majoring in aviation. It also gained additional state-of-the-art facilities in this key role for carrying out *ab initio* pilot training for Northwest Aerospace Training Corporation (Natco). This became fully operational by the end of 1989. The Natco operation in Eagan, Minnesota, began operation during January 1988 (Fig. 1-16). Most of the new simulators should be in place by 1990, and some of the older simulators were moved to Grand Forks so that training could begin there as planned.

Fig. 1-16. The Northwest Airlines Training Facility in Eagan, Minnesota.

Efforts are already being made to standardize the instructor's station in each flight simulator. A massive computer program is being constructed to do this, allowing instructors to be more valuable and flexible by being able to work in different simulator types. These programs are also designed to accept various types of training program information so when an outside airline uses the facilities (this is quite common), pilot candidates can be trained to any specific airline's qualifications and procedures.

As the existing simulators are standardized, other full-motion simulators will be implemented. Natco is spending $50 million for nine new Singer Link Airbus A320 and Boeing 747-400 simulators and cockpit procedures trainers (CPTs). (CPTs have no motion and are less expensive to operate than full-motion simulators. Each pilot spends time in these to become familiarized with an aircraft before training on the more expensive full-motion simulators.)

Besides contributing $3 million to build the new University of North Dakota facility, the FAA is also contributing $3 million for research. DC-9-10 and 727-100 full-motion simulators will be installed at the university's facility.

Northwest has committed to building these training facilities so that it can market its expertise at training pilots and tap the growing market for experienced pilots. Northwest claims that the number of airline transport pilots worldwide will grow from the current 107,000 to 400,000 in the year 2010. That's good news for those reading this book.

For now, hiring at Northwest will go on much as it has in the past. Natco will screen applicants and encourage those without college degrees to pursue the University of North Dakota's degree program. Northwest requires that sheepskin, and as long as enough qualified candidates keep coming through the door, that will be one of the airline's prerequisites. If qualified candidates become few, however, this qualification will be dropped, I assure you.

A great possibility exists that Northwest might fund some candidates if they are otherwise qualified to meet the needs of Northwest Airlines (the parent company). There are about 1,000 students enrolled at the University Center as aviation majors. This means that the University is already operating near its capacity, but as the new facility is completed, the *ab initio* program for pilots will move into full swing.

Many commuter or regional airlines are also interested in the program as they seek ways to train new candidates. Essentially, what is happening with this program is that the aviation equivalent of an American baseball team farm club is being contrived, where the regionals provide experienced pilots for their parent major airlines. The system will serve those wanting a top-notch aviation education and those that need top-notch pilot candidates. It's a good idea—as long as it is *not* used to break pilot unions using non-union pilots.

The British Airways Plan

While I've discussed what airlines such as Northwest are planning for the future in terms of pilot *ab initio* training and selection, the British have already embarked on a plan.

Over the past 20 years, British Airways has not grown or expanded very much. Recently some growth has begun, but more importantly, their pilot roster is beginning to show its age. The youngest pilot at British Airways is now 35 years old. Mandatory retirement age for airline pilots in Britain is 55. Thus, the British are planning ahead for a time when the pilot retirement schedule will be too high for training to keep up. British Airways is attacking the problem in two ways.

First, it makes sense to add pilots to the list right away, so the company has hired pilots (about 250) who are already rated with a minimum experience of 1,200 hours. Also, none chosen were over age 35. These first 250 pilots came from normal sources such as the military, air taxi operators, and from other airlines.

The second way that British Airways is finding candidates is through a school it has founded. This is a long-term plan to ease the pilot shortage. Eight companies bid on the project; British Aerospace won the contract and built a facility at Prestwick, Scotland. The facility is a flight academy in every sense of the term. It is much like a college with modern training facilities and dormitories for 128 trainees, plus some excess space devoted to training for other airlines.

Of most interest are the types of aircraft and training devices that will be used. The aircraft fleet will consist of Swiss FFA AS202 Bravo trainers, which have 180-horsepower Lycoming engines and side-by-side seating. The Bravos are also aerobatic because the British believe strongly in aerobatic training. The fleet will also include Piper Warriors and Seneca twins. A non-motion CPT based on the Warrior will also be used to facilitate training. Of particular value will be a full-motion, three-axis-of-motion Seneca simulator.

The first 50 hours of training will be done in the Bravo. Emphasis will be placed on basic aircraft handling skills, with some aerobatics added to the curriculum. Next comes instrument training in the Bendix/King avionics-equipped Piper Warriors and in the Warrior CPTs. After that, students will transition in a 45-hour program, 20 hours in the Seneca simulator and 25 in the Senecas. Finally, the students will learn about turbines by flying a bizjet simulator, that of a BAe 125-700. This phase also incorporates line-oriented flight training (LOFT), where students practice simulator airline flights with other students acting as crewmembers. During this phase, students will learn to adhere to British Airways procedures and checklists.

The British refer to this as "seamless" training, which provides training as close to the real thing as possible. The entire process takes 70 weeks and costs the airline about $80,000 for each graduate. Two British

Airways line pilots oversee the program and serve as mentors, role models, and observers of the operation.

Upon graduation, each student will have a Commercial license with an Instrument rating, as well as a passing grade on the ATP written exam. All that is lacking is a substantial amount of flight time, and that brings them to the next step—going to work for British Airways.

Even up to this point, a job has not been guaranteed, but presumably the pilot-graduates will go to work for British Airways. After all, the airline has just spent $80,000 per graduate, so it's into the right seat of a BAe 748, BAC 111, or Boeing 737 or 757. After a probationary period of two years (yuk!) at $20,000 per year (not bad), the pay comes up to speed.

If this sounds great, go for it. But the competition is stiff—*very* stiff. Some 8,800 people applied for the 128 slots. You've got to be sharp, level-headed, and have some natural aptitude for flying to make it through the battery of tests. Being a British citizen might help in this case, too.

Some other airlines besides Northwest and British Airways have full-fledged *ab initio* schools that build pilots from the ground up. Although these airlines don't normally hire U.S. pilots, they are Lufthansa, which trains pilots in Arizona, and Japan Airlines, which owns a training facility in Napa, California. United Airlines, as I've mentioned, is exploring its options and needs in this area, and in all likelihood will proceed with some sort of program. Eastern Airlines has had an arrangement with Dade County Junior College, where they look at students graduating from this program. This has neither the scope nor breadth of the Northwest or British Airways program and does not lead to a full college degree. Pilots should consider, however, that Eastern may not (at this writing) be the best airline to consider for a job, due to the convolutions through which Mr. Lorenzo is putting that once-proud airline.

How About Age?

The answer to this question is like some of the preceding questions: It really depends on the market. The younger you are when beginning this whole process, the better your chances will be. The best age is 22 to 28 years old. But please don't let that discourage you. I was 35 when I was finally hired and recently my cousin retired from the Air Force at age 42 and began his airline career with Southwest Airlines.

When hiring on after the age of 30, though, you may have to adjust your career expectations somewhat. Hiring on late can mean you will never achieve the most senior of positions—Captain of the largest airplanes may not be in the cards for you. This will vary, however, from airline to airline.

At least through the mid-1990s, don't let anyone tell you you're too old. It probably won't matter. In the next chapter, I'll discuss age and possible timetables for achieving your goals.

Conclusions

We've looked in a general way at what the outlook is for pilot hiring. The final chapter of this book will include actual statistics and more details on that subject. In this first chapter, I wanted to introduce you to the airline environment and help you understand that the future is bright, but that aviation has always had its labor problems. The benefits more often than not outweigh the problems we as pilots face in the industry. Couple that with the love of flying and there are few careers that can touch it. Because that is true, you can expect a fulfilling career.

Hiring prospects look good for the future, so now it's time to consider those items I discussed and determine the following: Do you want the airline *lifestyle?* Can you pass a First Class *physical?* Can you accept the idea of *two-tier wage scales?* Can you put the *time* and *money* into gaining the ratings and expertise you will need to become a true professional? If the answers to those questions are *all* yes, then go for it!

You probably had your mind made up before you bought this book. It helps, however, to have a true picture of what something is like before committing to a decision as hefty as this one. Many people told me I was crazy and should not pursue this career. I didn't listen, and except for details that change, I'm happy to say this career is everything I thought it would be.

If you have the qualifications and—most importantly—the desire, the goal is absolutely attainable. In the next chapter, I'll discuss how to formulate a plan of action and set goals. The next chapter is probably the most important in this book because a definite planned strategy will take you to your ultimate goal—an airline job—in the shortest, least expensive way possible.

I'm glad you're going to join me. I'll see you in the cockpit someday.

2

Building Experience and a Background

THERE ARE two things a prospective pilot should have before applying to even the first airline: a college degree and a logbook full of flight time. The flight time should better or equal most airline minimums. Now, it is not my intention to discourage anyone who might have all the flight time and no college degree from thinking he can obtain an airline position. A degree is an accessory, something the airlines like to see, although it doesn't directly relate to flying abilities or experience. When it comes time to fill a seat and totally qualified personnel are not available, the candidate without a degree will get the nod.

In this chapter, I'll primarily mention what airlines are seeking in qualified candidates. I'll help you design a career plan that will, in all probability, land you in the cockpit of an airline. But first, let's discuss this degree requirement a bit more and try to understand why it is desirable for a candidate to possess one.

It's a fact: Most people who obtain a college degree go further in life. They have higher incomes and are more often the ones who influence society in some way, either singly or collectively. It is also a fact that most college graduates are *thinkers*. This is the single most important ingredient for meeting airline expectations. By seeking degreed individuals, the airlines feel they are tapping the individuals who can think through the various aspects of flying and how these aspects affect each flight.

I think the airlines are onto something here. By and large, the pilots that I join in the cockpit are people who are highly intelligent and adapted to the job. The wonder of it all is why some airlines will reject candidates that other airlines will gladly accept. I have heard (although I can't substantiate it) that American Airlines looks at the *type* of degree a candidate has. It's been said that American prefers those who have degrees in the sciences or aviation itself. Liberal arts degrees, the rumor goes, are indicative of a personality that is "too passive" for the cockpit. The science degrees show an aptitude that meshes well in the arena of aviation. Indeed, the airline's attitude is that flying is a thinking man's game. *Of course it is.* Think of the times that a Captain must ascertain which way to navigate around weather for maximum safety and smoothest ride. This takes a complex thought process and a deep and varied background in more than a few disciplines. For example, meteorology, mechanics, aerodynamics, and fluid dynamics are all areas of knowledge that will at some time play an important part in the decision process of flying an airliner.

Out of these choices that the airlines make come an astonishing array of talented individuals. Indeed, it is rare to find a pilot who does not pursue some sort of outside interest. At Northwest, we have pilots who run the gamut from practicing doctors, lawyers, and dentists to ministers, mechanics, musicians, and real estate tycoons. Just about every field of endeavor is represented. What is the reason they occupy themselves apart from flying for the airline? The most obvious and important reason is that they tend to seek financial security and independence from the airline. This independence aids them during strikes, furloughs, or unexpected early retirement. The other reason is that flying for the airlines becomes somewhat boring. In fact, many feel it almost isolates them from the "real world" and therefore they seek to keep their interests in other occupations or hobbies for good mental health and a realistic attitude toward the world in general. If you are a neophyte and totally entranced with flying—as we all are when we begin—it might be difficult to relate to the idea of outside interests, but it is true and you will grow to understand it. If you are smart, you will start planning for it. Fundamentally, it revolves around having extra time on your hands to pursue other interests.

Now that you have an insight into the reasons for having a degree, it is up to you to decide how to tackle this hurdle. If you are a person who likes many things, then pursue a degree apart from aviation, or possibly a double major. Many years from now, you'll be glad you did. On the other hand, if aviation is an all-consuming passion, consider a program such as the University of North Dakota's. That way you might be able to achieve all your needed qualifications in one shot—that is, both flying and educational backgrounds all at once. Also, there are various scholarships and public grants that will help pay for your flying, as well as institutional learning at such universities. In the Appendix of this book is a list of institutions with aviation degree programs.

Airline Pilot Qualifications

In Chapter 1, I discussed qualifying yourself for the job, but actually talked more about what to expect than what actual pilot qualifications are. Before drawing up a plan of attack for your future, we need to start with basic requirements, especially flight time minimums.

Vision: Requirements are generally 20/20 uncorrected, but this has been relaxed in recent years due to the shortage of qualified pilots. Check into the particular airline you like before getting too involved in corrective measures.

Hearing: Hearing is very important as a qualification. Hearing loss cannot exceed 20 decibels in the frequencies of the spoken voice.

Blood Pressure: Average blood pressure must be below 140/90.

Height: Height should be between five feet six inches and six feet four inches. Some airlines have low limits of five feet seven inches.

Weight: Weight in indoor clothing without shoes should not exceed the maximum weight chart shown previously (Table 1-2). Some airlines (particularly American Airlines) will not budge a pound off this chart. (I know from firsthand experience. I have a very stocky build and at five feet ten inches, 174 pounds isn't even close to a good weight. I'm built just like a running back and am in excellent shape, yet I went over the limit and was rejected by American.) Here's a tip: If you're built heavy, empty your pockets before weighing and wear clothing made of a light (not heavy) fabric.

Flight Time: This is a hard item to nail down. I'll try to give you a general time bracket in just a minute, but first let me recommend the best aid in airline job hunting: FAPA, the Future Aviation Professionals of America. This organization is a profit-oriented company founded by Louis Smith, a Republic/Northwest pilot. The service is not cheap, but about 60 percent of all pilots hired by the airlines used this service last year. The current price is $168 for a year's subscription. This price will bring you all the current information on all airlines updated once a month for a year.

What does FAPA do for all that money? Basically, it provides you with current hiring information, such as which airlines are hiring how many pilots. This information is updated every month. When you first enroll you will receive a package that includes all the airline addresses and their minimum flight time qualifications. You will receive two kits, the Career Pilot Starter kit and the Pilot Career Information System. The information is extensive and up-to-date.

In *Piloting Careers* magazine, FAPA also provides enrollees with a counseling service that preps candidates for upcoming interviews with specific airlines. This counseling is tailored to the specific company in question and is invaluable when that great day finally comes.

To reach FAPA, just remember this number: 1-800-JET-JOBS.

Back to the subject at hand: Flight time varies from company to company. For a good commuter or regional airline job, 1,500 hours total time is a minimum. Of that 1,500 hours, you should possess 500 hours of multiengine pilot-in-command (PIC) time. Your actual instrument time should be somewhere in the neighborhood of 10 percent of your total flight time.

For that major airline job you're lusting after, 2,500 hours is a good minimum. Please understand, though, that these requirements move up or down according to supply and demand. These figures will stand you in good stead during times when pilots are bountiful. For the next five to seven years we can assume, if anything, the minimum requirements will move *down*. A good example is the newspaper ad I saw recently for Eastern Airlines. It was seeking only 1,800 hours total time.

Major airlines' demands for flight time are higher than those of regional airlines. In that same Eastern Airlines ad, the company required 500 hours of turbine multiengine flight time, where most regional airlines don't require turbine flight time at all (although most do require piston multiengine flight time). Whether piston or turbine, for a person just starting out in this business, 500 hours of multiengine time looks like an awful hill to climb. Don't worry; a carefully thought-out and executed plan will bring you to that airline interview with all the flight time you need. What's more, if you stick to your plan, a minimum of time will be needed to reach the goal of a job at a major airline.

Currently, the average minimum requirements for major airline flying positions are as listed in the chart below. If the minimums change, you'll know about it because you'll be keeping close track by being involved in the industry.

Total flight time:	2,500 hours
Multiengine (piston):	500 hours
Multiengine (turbine):	500 hours
Pilot-in-command:	1,800 hours
Actual instrument:	250 hours

These requirements are average for most airlines. In most logbooks you'll find other categories that you might want to fill out, but there are three that *don't* generally appeal to airlines:

Cross-country time is only important as you begin your flying career because it is required for various FAA licenses. Don't expect an airline interviewer to check at your logbook and say something like, ''Eighteen hundred cross country, huh?'' and then whistle like he's impressed. The point is, fill the column, get the time you need for your licenses, and then stop sweating it. Now, I didn't say never fill it out again; just don't worry how much time ultimately ends up in that column.

The second column that has little validity for professional pilots is night flight time. Again, the FAA does require a certain amount of night time for certain licenses, but it is not important for airline considerations except for displaying applicable recency of experience requirements for night currency to the FAA. Believe me, no one's going to care if you have 2,500 hours total time and 2,499 hours of it are at night—or just 20 hours, for that matter. Again, don't sweat the night number.

Third, there is a column for second-in-command (SIC). When flying for a regional airline, pilots usually start out in the right seat. That's normal. But in moving up the ladder, you must find your way into the left seat as quickly as possible. The point is, 2,000 hours of SIC time will look bad if your total is 2,500 hours. PIC is what's important to show you have what it takes to handle the job as a future airline Captain.

There is an exception to the SIC situation. A major airline will look most favorably on any SIC time that is *also* turbojet time. For example, 200 to 500 hours in the right seat of any airline-type airplane, such as a Boeing 727 or 737, will be looked on very favorably. This heavier airplane time will look better to the airline than right-seat corporate jet time—although that is good to have, too.

What About Flight Engineer Time?

There are two ways to go on Flight Engineer time. One is actual flight time, or at least a Flight Engineer Certificate. The other is possessing nothing more than a passing grade on the Flight Engineer's written exam. If you can get hired early on (with only moderate flying time under your belt) by, say, a freight carrier, then this time might be beneficial in obtaining a job. But the fact is that most major carriers only require the Flight Engineer written exam as a prerequisite to being hired. As fleets requiring three-man crews dwindle, then this aspect of getting hired may play a less important role in the future anyway, since two-man airplanes obviously eliminate the Flight Engineer's position. In any case, the bottom line here is simple: If you are offered a Flight Engineer's slot at a third-level carrier, take it! But if your ''light'' logbook needs to be beefed up with more hands-on flight time, then pass up the offer and concentrate on real flying. Flight Engineer time is just not that valuable in getting hired for flying jobs these days.

A Rating System

Because we just outlined what most air carriers are looking for in a candidate, let's try to put it in perspective. Many factors are involved, so for the uninitiated, a rating system will make the task of building flight time a little easier to understand. To put it simply, this rating system will help you ascertain what is important or valuable flight time. Also, you'll be able to determine when one particular flight time category tends to start losing significance and when other areas need to be bolstered so that you'll be able to present as rounded a picture of your experience as possible to interviewers. Each category is rated on a scale of 1 to 10 with 10 being the most important. Don't forget to consider the minimum requirements; these appear first in each category in Table 2-1. At first glance at Table 2-1, you might see some surprises. Don't worry; further explanation will clear up these seeming anomalies.

Table 2-1. Candidate Desirability Rating System.

Category	Rating or Desirability
Total Time	
*2,500 hours	10
2,500 - 4,000 hours	9
4,000 + hours	8
Multi-engine (piston) time:	
*500 hours	10
1,000 hours or more	6
Multiengine (turboprop) time:	
*500 hours	10
1,000 hours or more	9
Pilot-in-Command Time:	
*1,800 hours	10
2,500-4,000 hours	9
4,000 hours or more	9
Actual Instrument Time	
*250 hours actual instrument	10
10% or greater of total time	5
*Flight Engineer Written	10
Flight Engineer Certificate	9
Any Actual Flight Engineer Time	7
Jet Type Rating in any fleet equip.	7
Jet Type Rating not in fleet equip.	2
500 hours or less in turbojet	5
1,000 hours plus in a turbojet	8

*Indicates this meets the average minimum requirements for most airlines.

Table 2-1 is laid out in order of descending importance, the most important at the top and the least important on bottom. Some items can be transposed from time to time, such as when an airline is seeking particular requirements over others. In general, this chart should be normal for most airlines.

At first glance, the "total time" category looks somewhat confusing. While it is probably the one main requirement that will get an airline hiring department (not all airlines use their personnel departments to hire pilots) to notice an individual applicant, it is only the *minimum* requirement that needs to be met before others (such as jet time, Flight Engineer rating, etc.) start to take on importance. This is why the higher-hour time brackets have lower ratings. Another factor that enters into this argument is that once a pilot has proven his capability to fly professionally and has a good safety record, how much total time does this pilot actually need?

Next, look at the multiengine requirements. As you can see, 500 hours is the minimum. You can also see, however, that a great deal of multiengine time in piston-powered airplanes does not do much to bolster your appeal to an airline. The reason should be apparent: Airlines don't use piston equipment (with the exception of the smallest commuter or regional airlines—and if an airline is worthy of the term *regional*, I doubt that it has piston equipment at all).

From what I've just explained, it should be obvious that turbine-powered airplanes are the ticket to airline jobs. It's true. Most airlines use turbine-powered airplanes, albeit turbojets. And with the new ultra-high-bypass propfans that are being tested on jet airplanes such as the MD-80, who knows how highly turboprop experience will come to be regarded? (For the up-and-coming, hopefully, very highly indeed!)

As with the other categories, you can see that as turboprop experience goes up, its desirability rating goes *down*. The reason for this is simple: Major airlines fly pure jet equipment. Although a pilot might be gaining valuable experience with turboprops, that pilot is at the same time missing the experience of flying swept-wing airplanes. Turboprop time is great (and it's all *I* had when I was hired), but it reaches the point of diminishing returns after a while.

Pilot-in-command (PIC) time is incredibly important. In aviation, pilots must prove their capability. When a pilot has a good PIC record—ideally, accident-free—he becomes more appealing to any airline at any level. Getting as much PIC time as possible in the logbook is a most worthy goal. Notice that as the PIC time goes higher on the chart, the desirability factor remains high. While I hope that you achieve your goal of a major airline job before you log as many as 4,000 hours PIC, if you don't (for maybe industry economic reasons), you can see that it certainly won't count against you.

As far as instrument time is concerned, a good measure of actual instrument time is 10 percent of total logged flight time. This has been a rule-of-thumb in the aviation industry for as long as I have been around, and probably longer. At first, you should log all the actual IFR time you can fly. For one thing, it builds that confidence that all of us need to handle the strange environment of instrument flight. For another, it builds your actual usable experience level—that is, the sort of experience that you'll recall and need when the chips are down and the lives of your passengers depend on your correct action.

No matter what, a level of IFR time greater than 10 percent of total time is ho-hum to airline interviewers. The reason is that airline pilots fly little actual IFR. It's up and through it, clear on top, then down through it and out the bottom. Most approaches are visual. Since becoming employed at Republic/Northwest, my actual instrument time is probably less than *one* percent of total time.

One more warning about the 10-percent rule I just covered: A logbook with a lot of instrument time above that 10-percent level becomes suspect of P-51 (Parker P-51 pen) time. (That means hours logged that were never actually flown—doctored logbooks, an unfortunately common problem in aviation.) If you fly more than that 10 percent level, log it; then challenge the airline to prove you *didn't*. I repeat: *Don't* add instrument time to reach more than about that ideal 10-percent level, thinking it will make you look good. It will have the *opposite* effect. And don't log instrument time you haven't flown because, believe me, lack of true experience will show up on your first simulator check ride. When taking a simulator check for an interview, you will most likely never have flown that type of airplane before. The only advantage you will bring to that unfamiliar simulator is your actual instrument flying experience. Make sure that experience is real.

Next in importance to the airline is the Flight Engineer requirement. Every passing year this will become less important to the majors as the two-man crew concept becomes entrenched with new-technology airplanes. You can see that actual third-seat flight time will be of little value, especially when the airline that hires you will necessarily have to train you to their standard procedures anyway.

Bear in mind that although the major airlines are limiting their three-man cockpit crews, the freight haulers are *adding* to their fleets a number of these outdated airplanes. Federal Express and UPS are among those that still find the three-manners are profitable for carrying freight and good investments due to low purchase price. Expansion in these areas is possible.

The jet type rating has been a big question mark for aspiring airline pilots for a long time. To buy a jet type rating on your own requires lots of money. Most pilots (like me) could never afford a jet type rating regardless of the advantages. Some airlines (Southwest, for one) require a type rating. But I have to admit, it is a crapshoot. The airline that hires you might not even have a jet that matches the type rating for which you spent so many dollars. Therefore, that explains the poor showing that type ratings get in the desirability ratings. It's your choice.

Finally, a type rating can be worth a little more when it is accompanied by some flight time in the aircraft. This requires, however, that you pick up the flight time by being employed somewhere. Any jet job is a good job, so you might have to look closely before you leap to something that may or may not be that much better. Remember seniority.

I haven't forgotten that jet time can be acquired without the type rating. There are copilots jobs out there, and experience in a jet—particularly of a type that an airline already has—is worth much to you and will assure you of strong consideration for that job. In the overall scheme of things, the best qualified pilot will be able to pick and choose when he has jet experience, not just a type rating. One thousand hours in a jet and any airline will think you are ''something special in the air.''

Planning the Steps to Your Future

Here we are at the point that will make all the difference to you in the future. We've already discussed how flight time is evaluated by airlines. With that knowledge, you already know more than I did when I started. This next step will make your trip to the ultimate goal—airline new-hire—as easy as possible. Efficiency is also part of the plan, because nobody wants to expend unnecessary money, time, or effort reaching this goal.

Setting goals is important, and that's the next subject. There is one attribute that will be more important than any other in reaching your ultimate goal, and that item is *desire*. Many obstacles stood in my way to an airline career, as I mentioned in this book's Introduction, but one thing should be apparent from my story: Desire will carry you through every adversity.

Many people might suggest that determination is what carries one through adversity, but I don't agree. Without that deep desire, there would be no determination. What lights that determined fire, I cannot say. But what *keeps* it burning is desire.

So how do you know if you have that sort of desire? For one thing, the person who has it *knows it*. It's a pretty good bet that at least early on in the experience you will have a voracious appetite to learn everything about airplanes and flying them. If your desire to fly is utmost in your mind, you'll go to sleep thinking about flying and wake up thinking about flying.

Setting Goals

I've read several books about setting goals and they all say the same thing. Ultimately, they conclude by saying that setting goals is a simple matter, and how they get a book-length dissertation out of such a simple subject is beyond me. I've found that most people don't know how to set goals. Here's what the books say:

First, I feel that we have established that you want a career as an airline pilot. That is the final objective. So far, so good. But if you have only that goal in mind, you may never attain it because of the discouragement that you will certainly face from time to time in pursuit of that goal. The key, then, is to set *intermediate* goals. These intermediate goals are mileposts that gauge the progress you are making along the road to your goal.

What should your intermediate goal be? It is impossible for me to choose the exact path for thousands of readers seeking the "ultimate goal." But as we have discussed, there are two main ways of getting there—the military route and the civilian route. Because the civilian route is much more complex and is also what I know best, I'll discuss that first. Then I'll chart the goals in the military direction.

If you are considering the civilian route, consider these two thoughts: It is expensive, and it doesn't take as long to accomplish as the military route to an airline job. It's a tradeoff that we have already explored earlier in this book. So let's assume now that you have chosen the civilian training route as I did. What *is* that route on the average?

Many thousands of hopeful airline pilots will read this book. Each one is different, and the circumstances they meet will be very different. From research I've conducted over the years, you can safely assume that the process is much the same for everyone. Here it is:

1. Interest begins.
2. Research subject.
3. Take the first plane ride.
4. Begin training for pilot's licenses.
5. Flight instruct to gain flight time and experience.
6. During the instructing phase, gain as much multiengine flight time as possible.
7. Get a job as a charter, freight, or commuter pilot.
8. Transition to Captain of a regional airline turboprop or jet, or corporate jet.
9. Get hired by a major airline.
10. Pass the probationary year with flying colors and finally rest assured that there is a future at this airline.

This ten-step process is representative of short-term or intermediate goals. These intermediate goals are important because they let you observe your progress as well as gauge when to move on to the next part of the plan.

Keep in mind, however, that these steps are generalities. It would be best to construct a plan with mileposts clearly displayed, so let's try that attack. The first three general steps in the above plan can be done at any time, but once you are into Step 4, the plan is definitely underway.

Once more, I want to reiterate that college is most beneficial if not mandatory for anyone preparing for an airline career. The reasons were explained earlier. But now you'll have to decide whether you will learn to fly as a part of college, during college while pursuing a major outside of aviation, or after college is finished and money for gaining licenses is more available. Because these separate plans of attack are so different, I'll just discuss the plan as if you're already into the plan and try to identify the mileposts to which I keep referring.

Before I get too deep into this career plan, I'd like to address those of you who might be quite young and want an early start. See Table 2-2. There are ways to do this. If you are a sophomore in high school, here's a tip for you: You can begin flight instruction for a Private Pilot's Certificate at age 15, solo at age 16, and take a check ride for the license at age 17. This tip is that many FBOs (fixed-base operators, companies that offer services at airports such as fuel, flight instruction, charter, etc.) need people to help run their business. FBOs will often hire young people to wash and fuel airplanes and help in other areas in exchange for flying time. The benefits are obvious if you are young and interested in an aviation career. If this interests you, hit up all the local airports in your area and ask. Just hanging around the airport can be fun and educational and could result in the start to a great career.

If you are a little older and ready to get into the real world of flying for a living, this is the best way to go: Select the flight school for your training and apply yourself to the program as much as you possibly can, given your financial situation. A flight school operating under Part 141 of the FAA regulations can get you through the quickest. Get to know all the people who run the school so that when you graduate, you can ask them for a job as a qualified Flight Instructor and later as a charter pilot.

I should point out that the Airline Transport Pilot (ATP) Certificate is required by some employers. Others require just the ATP written. Many airlines will "supply" their pilots with an ATP when the pilot first upgrades from First Officer to Captain. Because this varies so much, I have not attempted to indicate when you should acquire the ATP. Consult FAPA and look at your own circumstances, then decide if the money will have to come out of your own pocket.

According to Step 4 of the general plan, you must acquire the Commercial/Instrument tickets to become semi-employable. At this stage, with the right amount of flight time, you could fly single-engine VFR charters (not likely) or fly pipeline patrol, tow banners or gliders, or fly parachute jumpers. Because of the little most pilots can offer to prospective employers at this point, most elect to pick up additional certificates

Table 2-2. Minimum Requirements for Certificates.

Student Pilot

Minimum Age: 16

Class of FAA Physical: 3rd

Flight Experience: None

Private Pilot

Minimum Age: 17

Class of FAA Physical: 3rd

Flight Experience per Part 61:

Total	Dual	Solo	Instrument Dual	Total Inst.	X-C	X-C Command
40	20	20	None	None	12	10
		Night: 3				

Commercial Pilot

Minimum Age: 18

Class of FAA Physical: 2nd

Flight Experience per Part 61:

Total	Dual	Solo	Instrument Dual	Total Inst.	X-C	X-C Command
250	50	100	20	20	53	50
		Night: 5				

Airline Transport Pilot

Minimum Age: 23

Class of FAA Physical: 1st

Flight Experience per Part 61:

Total	Dual	Solo	Instrument Dual	Total Inst.	X-C	X-C Command
1,500	50	250	20	75	500	50
		Night: 100				

X-C means cross-country.
Dual means with an instructor.

such as Certificated Flight Instructor (CFI) and Flight Instructor Instruments (CFII), plus the Multiengine rating and Multiengine Instructor (MEI). Most flight schools don't usually let fresh instructors teach multiengine, but there are some that do, and in any case having the ratings will be a start and will encourage the boss to throw multiengine crumbs your way.

With the above certificates and ratings in hand, you can cross off Step 4 and act upon Step 5. Making this step is the first important milestone in your now-burgeoning career.

Many pilots think of Step 5 in one of two ways: "I don't want to do this," and "I don't want to do this for long." Thinking either way can be a mistake, depending on the type of person you are and the opportunities that come your way. If you can avoid teaching flying and proceed to the next step, good for you, if you don't like the idea of teaching. On the other hand, teaching flying will provide two benefits for aspiring airline pilots: It will build your confidence in handling airplanes that have been put in dire circumstances by ham-handed club-footed students, and each time you return from a lesson, your log will have grown at least another hour of PIC time. Both benefits are very important.

Another side benefit of this is the passing on of good techniques to others who trust your judgment and ability as a pilot. Some of your students may go on to become airline pilots themselves and the aviation industry is an extremely small world, so teach them well!

How long will you have to teach? That depends on two things—total flight time and multiengine time. One thousand hours total time is a good figure to have in your logbook before proceeding to Step 7. But don't forget Step 6, which is the second thing you'll need to accomplish before leaving Step 5 behind. Even though 1,000 hours might be enough to proceed with Step 7, there must be at the same time some multiengine time in your logbook. Acquiring 300 to 500 hours multiengine time before Step 7 will be sufficient for you to find a full-time multiengine flying job that will propel you further along your path.

One of the hardest questions you'll face making your career goals come true will be logging multiengine flight time. How do you get multi time? Believe me, it comes slowly, but here are some ideas. Lots of FBOs fly multiengine charters. If you teach at one of these FBOs, after you get to know the right people, ask to fly right seat on some of the charters. FAA rules in this regard are very strict, so check these out before you ask. You can't log time unless you have a multiengine rating and are the sole manipulator of the controls, so unless you are flying the airplane, you can't log the time as PIC. Unfortunately, you can't be sole PIC when the actual charter flight is underway unless you are authorized as a pilot for that flight, so you can only log time (fly the airplane) on a deadhead or return leg that isn't part of the actual charter flight. A side benefit of this type of flying is that you are getting supervision from a seasoned professional. It's a time-honored method of logging multi time, but there is a lot of competition for those empty right seats, so do what you can to be invited along.

Another idea is to make friends with some of the smaller local freight haulers or medical charter companies. They often run airplanes as two-crew operations and need qualified right-seat pilots, in which case the right-seat pilot can legally log the flight time as SIC time. Some of these companies even pay right-seat pilots.

Some multiengine flight instruction might come your way, and this is one of the best ways to log a lot of multiengine time quickly, especially if you are connected with a busy Part 141 flight school that specializes in training professional pilots and has a large fleet of twins.

The main point—and this is *extremely* important—is that you need to seek out every available opportunity to acquire multiengine flight time. Completion of Steps 5 and 6 should coincide with your obtaining the minimum amount of multi time.

At Step 7 you begin to take on the swagger of an old salt. Flying has started to become the way of life to which you've become accustomed (Fig. 2-1). If you're lucky enough to sign on with a commuter airline, you'll be flowing in the mainstream of what is happening in the airline industry. Although I started Step 7 as a freight pilot, it is much easier for an up-and-coming professional to find his way into a commuter or regional airline these days with relatively low flight hours logged (Fig. 2-2).

This is a good opportunity to clue you in to some recent positive events. In a first for the airline industry, Ransome Airlines, a long-established regional carrier, signed an agreement with Pan Am that makes new-hire Ransome pilots an integral part of Pan Am's seniority list. This has the effect of stemming the high turnover rate at Ransome and assuring Pan Am of a source of dependable, qualified pilots. It's a win/win situation for everyone involved.

Fig. 2-1. Cargo operators are good places to gain experience. Some, such as this one which is contracted to Federal Express, offer a pilot a chance to gain valuable turboprop experience.

Fig. 2-2. It doesn't take long to move up to Captain in one of these light turboprops—and it's good experience!

Northwest has a similar type of situation. Northwest feeder Express One has signed an agreement that puts its pilots on an evaluation program with Natco (mentioned earlier). Although this isn't the same as the Ransome/Pan Am agreement, it tends to have the same results. In this case, it allows Northwest to pass final judgment on qualified candidates.

Knowing when to leave Step 7 for Step 8 is just a natural process. If you happen to be flying a corporate jet, the opportunity may not arrive as quickly as it might if you are flying for a regional airline. The reason is that attrition among corporate Captains is much slower and corporations don't often expand their flight departments, either. It is something you should consider before accepting a position at a corporation if you ultimately want an airline career.

Attrition is very high at regional careers and they generally operate on a seniority basis, as do the major airlines. Therefore, what I am saying is that moving from Step 7 to Step 8 is a very natural process in an airline environment.

Step 8 to Step 9 is the most difficult process in this plan. The step discussed earlier regarding Ransome/Pan Am and Express One/Northwest or regionals formally aligned with major airlines might make the step much easier.

Knowing when to make the leap will be very apparent when you arrive at this point in your career. For planning purposes, I've already identified that point: It is the point at which you arrive at the minimum requirements of the airline to which you are applying. Any extra flight time or experience as a check airman, assistant chief pilot, or even chief pilot will make you that much more desirable to major airlines (Fig. 2-3).

Fig. 2-3. The big time! This is a Continental DC-10.

Step 10 is accomplished on the anniversary date of your hiring by the major airline of your choice. Getting there from the first day of class is fun, scary in an uncertain sort of way, and challenging. And until that probationary period is complete, you cannot and will not rest easy. In Chapter 5, I'll discuss the entire experience of probation.

FlightSafety's Airline Transition Program

Let's backtrack for a moment to Steps 5 and 6 and look at a program that could prove to you to be a short cut to Step 7.

Long a mainstay for professional instruction in the corporate aviation industry, FlightSafety has now developed a program aimed at helping the up-and-coming airline pilot. Not only does this program help the aspiring pilot, it also helps the regional airlines find qualified personnel. Here's how it works: FlightSafety's Airline Transition Program is a 12$\frac{1}{2}$-day course designed to prepare the recent flight school graduate for a position as a First Officer for regional and commuter airlines. In other words, pilots with a minimum of flight time—say, 250 hours, for instance—can enter the program. What this can do is allow the individual to bypass the usual ladder of building flight time to attain the necessary experience to find a commuter/regional job flying.

The course includes 75 hours of academic instruction and 25 hours of simulator instruction, including 12.5 hours as flying pilot and 12.5 hours as non-flying pilot. The course is not aircraft-specific, but is designed to introduce the pilot to pressurized, turbine-engine aircraft and the environment in which they operate.

It's easy to understand that the typical flight school graduate spends most of his time in general aviation equipment. FlightSafety exposes the prospect to advanced systems such as pressurization, hydraulics, pneumatics, and turbine engines. Since cockpits have changed greatly in the last few years, EFIS (glass cockpit, television displays), INC (inertial navigation systems), and flight director systems are also introduced. Airlines use Part 121 and 135 of the Federal Aviation Regulations. These regs, along with weather, weather radar, operation specifications (known as the ''airline's bible'' because it is FAA-approved), and crew concept and cockpit management are addressed in the ground school.

Before a trainee is admitted to this course, he is prescreened as to aptitude to aviation and psychological suitability. (In other words, it helps to be crazy!) FlightSafety's prerequisites for enrollment into the Airline Transition Program are for trainees to hold a Commercial Pilot Certificate with multiengine and instrument ratings. In addition to possessing 250 hours of flight time, a First Class Medical Certificate is also necessary. As you can see, the requirements are right in line with what a pilot earns in most commercial pilot programs.

Sometime prior to or following FlightSafety's prescreening, a regional airline will give the candidate a prehire interview. Depending on how well the applicant does at both the prescreening and interview, he will then be selected and admitted to the course. The idea here is that the applicant may be hired by that regional airline upon successful completion of the Airline Transition Program. Upon employment, the trainee is sent through ''type-specific'' training conducted either by the regional airline or by FlightSafety. The end result is a job as First Officer for the selecting airline.

All of this requires the pilot applicant to put up a great deal of money. Essentially, you should think of it as investing a little more money in your career in order to short-circuit the normal ladder-climbing process. We'll discuss how much it costs later, but we should go over this system again to make sure you understand how it works.

First, the entire process requires a $300 nonrefundable fee up-front. For that money, you get evaluated (simulator flight, checking instrument proficiency), prescreened, and interviewed by a regional carrier participating in FlightSafety's program. Only *after* the airline has agreed to hire you (if your final check ride is successful in ''type-specific'' equipment) must you put any more money into the program. This protects the applicant from posting an inordinate amount of money to secure a job that might not be a sure thing. In other words, only when a job is stipulated by a successful check ride (in which the chief pilot observes your ride) is it necessary to put out any more money over the initial $300.

The entire cost depends upon aircraft type. FlightSafety says to expect to pay between $5,100 and $7,000. It's not a small amount, but it may save you years of flight instructing or flying night freight in dubious aircraft to achieve the same thing—a job with an airline.

The Military Route and Plan of Action

As you know, I went the civilian direction. I could have gone the military route, but I really didn't know that was available to me until it was too late—but that's another story. I do know enough of the process to advise you here because of my numerous acquaintances at various airlines who came up the military ladder. What follows is what I have gleaned from their experiences.

To fly in the military, a college degree is necessary. It is a prerequisite for gaining the needed status of 2nd Lieutenant. There are two short cuts, though, that can help you if you decide to go the military route, and both involve learning to fly earlier. Many colleges offer Air Force ROTC and some ROTC candidates are trained in primary flying skills through local flight schools at government expense in exchange for a commitment to serve after graduation.

The other short cut is to qualify for the Air Force Academy. Primary flight training is part of its curriculum. But be warned: There is a great deal of competition to get into the Academy.

As a basis for discussion, let's assume that if the military route interests you, you already have completed your college education. Then the questions are: Which branch should I choose? What's the plan? Let's take a look at it.

The military is just not turning out the number of pilots it used to. The reason for this is that incentives are higher now for pilots to reenlist. With airlines hiring at B-scale levels, many military pilots' salaries now rival and exceed B-scale rates of pay. Many pilots see the military as security—stay in for 20 years, earn a decent government pension, and then, if there is demand, take a swing at airline flying or some other occupation.

On the other hand, many military pilots can't wait to do their six-year stint and get on with life outside the military. The military life requires much moving, and in the Navy, for instance, pilots are frequently separated physically from land for months at a time. If a pilot has a family, he may not want to risk being absent for six to seven months at a time. To me, that would be too heavy a price to pay for learning to fly. There are other opportunities that don't extract such a high price.

If you have your mind made up that you will join the military, you do have to pick only one branch to join. How do you choose? It depends on what you want: Shortest enlistment? Best training? Best duty?

In all branches except the Army, a bachelor's degree will be required. The U.S. Army will settle for a high school diploma. You will need 20/20 uncorrected vision, and that requirement will shoot some of you down right there, although it will not necessarily keep you out of the airline profession. Also, keep in mind that the military is accepting women now for a greater variety of assignments than ever.

U.S. Army: The least restrictive of the military services. Ages 17 to 35. All flight training is for helicopters. Fixed-wing training can be obtained after helicopter training. However, the availability of fixed-wing positions is small and jet positions are almost nonexistent. The stay is for four years after obtaining your wings.

U.S. Air Force: Probably the most beneficial training and positions with respect to future airline careers. Ages 18 to 27.5, but you must apply before your 26th birthday. The catch: seven years of duty to complete *after* earning your wings.

U.S. Marine Corps: Ages 18 to 27.5. A four-year stint is mandatory.

U.S. Navy: Ages 18 to 27.5. Five-year enlistment after training. Some of the most challenging and exciting flying in the world.

National Guard: A limited number of pilots are trained for entry directly into the National Guard. They are trained by the respective branch in which they will serve, either Army or Air Force.

Reserves: Not many pilots get a chance to go this way initially. Usually one must have served his active duty commitment before going to a Reserve unit. But the Reserves are a haven for junior airline pilots who might be flying as Flight Engineers. Their only hands-on flying during their Flight Engineer days is usually at the Reserves. It also provides a nice secondary income during those early low-paying years. I have noticed that the guys who fly in the Reserves during probation usually do much better financially than others.

So how do you decide? That's up to you. You might have a preference due to family tradition. You might just want the lowest service commitment. Whatever, yours will be the final decision. Once you decide, then you will know the minimum amount of time it will take to reach that airline job.

After the Military

Most military pilots come out of the service with little or no idea of what is happening in civil aviation. To remedy that, you should subscribe to FAPA. All the latest information will be provided to you. Then start attending interviews and talk to other applicants. Shortly, you'll know what's going on in the civil aviation world.

Before that first airline interview, there is one important item that must be taken care of and that is the conversion of military pilot certificates to civilian ones. Generally, this involves taking dual instruction for the ATP license and taking a check ride with an FAA inspector. Once this is done, you will be ready to pursue that coveted airline career. Within a month of discharge, you should be in the hiring pipeline—and a very desirable candidate, at that.

One thing I didn't mention yet: What sort of flight time do military pilots usually acquire during their service commitments? This varies from individual to individual and also hinges upon length of service. The *least* I have ever heard of was 1,800 hours. The highest amount logged by those who were in only one enlistment was about 4,000 hours. Thus, between 2,000 and 3,000 hours is a good average. This is commensurate with most airline minimum requirements. Most of it is jet time and therefore will be more readily sought after by the airlines than an equivalent amount of civilian turboprop flight time.

There you have it—the way airlines evaluate experience. You have two major choices, civilian or military, and we have discussed drawbacks and advantages to both routes. One thing is sure: A positive plan is here for you to follow. It is a matter of finding the shortest distance between two points and setting intermediate goals. This is how it's done.

Next, we'll discuss how to actually go about getting that job. I'll show you how it's done and how to get those interviews and lock that job up tight.

3

Getting the
Airlines' Attention

TO THIS point, the trip to that airline job has been a challenge. By now you have honed your ingenuity to a fine edge and logged hours of valuable flight time. You've got hours in your logbook in flying machines you never knew existed. An airline job is all you've thought about for years, and your strong desire has pulled you through all the hardships and hard knocks. Now comes the time to assert yourself and attract the attention of the people who hire pilots. You have to stand out from the rest of the applicants— but in just the *right* way. Determination might not be enough. Here's where a few tricks of the trade will be useful—even crucial. That's what this chapter is about. I'll discuss the procedure for getting that all-important interview.

Double-Check Your Qualifications

I've already mentioned how important FAPA's help is. At this stage, FAPA becomes *very* important. Also, I assume (and hope) that any pilot who has reached this point in his career has been asking any and every airline pilot he knows about which airlines are hiring and who got hired with what sort of qualifications. FAPA gives statistical averages on latest new-hire qualifications. This will be most helpful in sizing up the competition. Knowing friends who get hired and having first-hand knowledge of their qualifications will help you to determine if you are ready to go after that airline job. It is very encouraging if you know that you are the equal of people who have been hired because *your* job might very well be just around the corner.

To reiterate: Check the latest FAPA statistics (generally found in FAPA's monthly magazine). Keep your ear to the ground and know what sort of qualifications your friends and acquaintances are holding when *they* get hired. If you are near being competitively qualified, get started with the following step.

Preparing a Resume

A resume in job hunting—especially in the airline business—is used in lieu of an application. Applications can be obtained, but it is a good idea to send a resume when writing for an application. The reason for this is that your resume immediately creates a file in your name at the airline. Also, if a hiring rush is on, the airline might be so short of pilots that it may extend an invitation for an interview even before you return the completed application. Ordinarily, the idea is to get hired *the first good day that comes along*. Remember seniority?

Resumes can be done on your own. Shown in Fig. 3-1 is the one I used to get hired at Republic Airlines. Follow it as a guide, emulate it, use it any way that you want. Just make sure yours looks as good as, if not better than, this one.

Having a good-looking resume does not require that it be professionally prepared. If you copy my style and format, you will have a good, solid, interview-producing resume of which you can be proud. All you need is a decent local printer and an attractive heavy bond paper. My paper was an off-white sort of beige. I selected this color as more pleasing to the eye than stark white, yet I felt it showed individuality with a touch of class without making a loud declaration. A point to remember: Airline companies want their candidates to be very conservative—clone-like almost—with just one or two interesting attributes to distinguish them from the pack.

There are many good reference books on preparing resumes. If you check, you'll find that my resume is much like the norm, so I won't spend any time explaining how to prepare one. Just use mine as a guide and fill in your own numbers; that's all you need. Also, you should know that FAPA offers a resume preparation service where FAPA keeps your statistics and numbers on file so that you can obtain from FAPA an attractive, up-to-date resume whenever you need one.

Once you have your freshly printed resume in hand, it's time to send some out. Each one should be accompanied by a short query letter. Normally, these letters should be typed up for each individual airline. A form letter might do, but I discourage these because they lack a tailored-just-for-that-airline appearance. Airlines tend to be very egotistical and want to believe that you took the time to write them because they are at the top of your list—even if *you* know that is not the case.

An example of a tailored comment is as follows: "I would value the chance to fly for American Airlines and carry on the family tradition that my father started 35 years ago when he was hired by American." Now please don't feel intimidated if no one in your family ever worked for an

Present Address 601 Chaucer Way, Lawton, OK 73505
Permanent Address Box 6391, Lawton, OK 73506-0391

Objective:	Pilot or Flight Crew Position			
Flight Time:	**Total Time**	**7,260**	Turboprop	5,321
	Pilot in Command	6,657	Multi-Engine	362
	Second in Command	450	Simulator	77
	Instructor Pilot	798	Instrument	443

Certification: **Airline Transport Pilot** (SD3-30, CV-580 types)
Commercial Single Engine Land
Certified Flight Instructor (Airplane & Instruments)
Flight Engineering Written (FEX) Expire 2-87
FAA Class I Medical No restrictions

Education: Nederland High School, Nederland, Texas 77627 June 1968
Bachelor of Science in Geology, Lamar University,
Beaumont, Texas   Graduated December 1972

Employment: **October 1977 to Present**

Metro Airlines, Inc.
8505 Freeport Parkway, Suite 600
Irving, Texas 75063
(214) 594-3400
**Have been serving as Captain CV-580
for the past four years.**

March 1977 to September 1977

Great Western Airlines
Formerly of Tulsa (now defunct)
Flew as First Officer then **Captain on
Beech 18 Turbine.** Flying night freight.

April 1976 to March 1977

Professional Aviation Services
Formerly of Nederland, Texas (now defunct)
Flew as flight instructor and charter
pilot.

**Additional
Information:** **I have written seven aviation books** of a technical nature. They deal with flying in the **IFR environment, weather forecasting** and the humorous "Hangar Tales & War Stories" "Instrument Flying" made the publisher's best seller list.

Also, I have been **operating my own video production company** in which we have been **producing flight training tapes on the CV-580** for Metro Airlines.

Personal: Date of Birth: February 21, 1950
Weight: 175 lbs.
Height: 5' 10"
Marital Status: Married with one daughter
Health: Excellent/Non-Smoker
Hobbies: Photography, Writing, Skiing and Hiking

Available Immediately, Will relocate, References on request.

Fig. 3-1. Author's personal resume at the time of being hired.

January 19, 1990

Mr. Art Davis—Personnel Manager
Delta Airlines
Employment Office
P.O. Box 20530
Atlanta, Georgia 30320

Dear Mr. Davis:

In recent days I have completed my training and have obtained my Airline Transport Rating. In addition, I have completed my Flight Engineer Written Exam. I feel this now completes my qualifications and brings me to a level that can be considered for employment by Delta.

Enclosed is a current resume' of my pilot qualifications for your review. I also would like an application from Delta in order to receive consideration from your department as a pilot applicant. Delta has always been at the top of my list and now that the time has arrived that I can apply to Delta and be competitive is a dream come true. I thank you for your time and attention to this letter and hope to visit with you or your staff in the near future.

Sincerely,

Jeff W. Griffin

Fig. 3-2. Sample query letter.

airline, or has ever even ridden on an airplane. This is just an *example* of a tailored comment. Every applicant for an airline job should put at least one such comment, relating to the applicant's own background, in each query letter sent out.

Once again, follow the form of my query letter (Fig. 3-2) and you'll be fine. Just change each one ever so slightly.

Recommendations

In most businesses, recommendations are invaluable aids to getting hired. In the airline business, a recommendation can get you the job, or it can be the kiss of death. As a general rule, submit recommendations *only* to companies that are known to solicit them. Southwest Airline, for example, requires two Captains to recommend each candidate. Delta Airlines, on the other hand, seems to place little or no importance on recommendations—or so the stories go.

For the moment, let's assume you live in a small town—perhaps you are female—and have no friends or relatives in the airline business. In other words, the odds are really against you. How do you go about getting that recommendation? Obviously it won't be a snap, but here are some pointers:

As simple as this sounds, airline pilots can be found at airports. They also fly regularly scheduled flight patterns. That will generally put them in the same place at the same time every week for one month at a time. Let's take the small-town scenario a bit further. Let's say you have made your way to being a copilot on a regional airline. Next, you see a Captain from a major airline commuting home on your flight. Chances are that next week at the same time he'll show up again. This is your opportunity to get to know him. Once you do, he'll recommend you—simple!

Another way to find recommendations is by getting to know the other pilots with whom you fly. Many have fathers, brothers, sisters, or more distant relatives who fly for other carriers. Approach them and see if this avenue might be open for a recommendation. In other words, *be resourceful*. Don't be pushy; ask politely. No one will turn you down if you know them somewhat prior to asking for a recommendation (Fig. 3-3).

In any case, the lack of a recommendation is no handicap unless the airline specifically requires it prior to granting an interview.

The How and Why of Applications

Applications are a necessary part of applying for any job. You can consider them as an opportunity to ''blow your own horn.'' As with any job, negative information doesn't do any good; therefore, avoid putting negative information on job applications. Airlines are even more sensitive than most prospective employers. It seems airlines give little leeway

Continental Airlines

Fig. 3-3. Do you know the Captain or the First Officer of this Continental 737? If so, their recommendation could help you land an airline job of your own.

55

to applicants, so when filling out an airline application you must take extra care and attention to put your background in the best possible light.

Shown in Fig. 3-4 is a sample application (from the fictitious "Big Wings Airlines") for your consideration and analysis. I won't insult your intelligence by implying you don't know anything about job applications, but I would like to mention a few points about this task to make it easier for you.

In the first section—"Personal Data"—you are asked to list all the people you know working for the company. Before you answer this question, do some research to find out if the company supports or rejects "nepotism" or the hiring of close relatives. The way to answer this question is to leave relatives *with a last name different from yours* (such as cousins, nieces/nephews, uncles, etc.) *off* the application. The company cannot in any way spot these relations without your help. Some say this is dishonest. Maybe, maybe not. To me, what is important is that with deregulation the number of airlines has dwindled significantly. This means eliminating yourself because a relative works at the same airline cuts out a number of possible job offers even before you get started. After all, what a cousin of yours does somewhere else in the company will have little bearing on your own success.

If, on the other hand, your father or mother or brother/sister or aunt/uncle *with the same last name as you* works for an airline, it will be next to impossible for you to plead ignorance if the airline discovers the relation and you have neglected to mention it on the application. Be up-front with those names and accept the fact that with some companies there's a good chance you won't even be called for an interview.

Personal acquaintances can be a help or can hinder you in getting hired. Do a little research before putting someone's name on that application. All airlines keep files on employees that are more like FBI dossiers than friendly records. This is so that successive changes in management will be able to recall specific instances when things went well or badly (*especially* badly) for the employee. Rest assured, they will check anyone's name that appears on an application.

Under the heading "Position Information" there are general questions that will disqualify you if you don't answer with a positive or affirmative answer. For example: "State your reason for interest in our company." An answer like "Well, you guys are the only company that hasn't turned me down" won't do you much good at all. So what should you write? That's easy: Address the company's ego. Try something like this, for instance: "I'd like to work for Northwest because of its strong financial base which promises job security, and I'd like to do my part to uphold the tradition of having the finest-trained pilots in the industry."

The above statement can be applied to almost all major carriers with the exceptions (at this writing) of Eastern, Continental, and Pan Am.

IMPORTANT
READ CAREFULLY

Print or type answers. Answer all questions except those prohibited by applicable State law.

The Age Discrimination in Employment Act of 1967 prohibits discrimination on the basis of age, with respect to individuals who are at least 40 but less than 70 years of age.

BIG WINGS AIRLINES

AN EQUAL OPPORTUNITY EMPLOYER

APPLICATION FOR EMPLOYMENT
This application for use by **Pilot** applicants only

RETURN TO: BIG WINGS AIRLINES, BOX 1234
MONEYTOWN, USA 12345

PERSONAL DATA

FIRST NAME	MIDDLE	LAST	DATE
PRESENT ADDRESS IN FULL	CITY	STATE / ZIP	TELEPHONE ( )
PERMANENT ADDRESS IN FULL	CITY	STATE / ZIP	TELEPHONE ( )
SOCIAL SECURITY NUMBER	DATE OF BIRTH	HEIGHT	WEIGHT

ARE YOU A U.S. CITIZEN? ☐ YES ☐ NO	TYPE OF VISA?	ALIEN REGISTRATION NO.	HOW WERE YOU REFERRED TO BIG WINGS?	ARE YOU A DESIGNATED EMPLOYEE PER SECTION 43 OF THE AIRLINE DEREGULATION ACT OF 1978? ☐ YES ☐ NO

LIST RELATIVES, INCLUDING IN-LAWS (AND RELATIONSHIP) AND PERSONAL ACQUAINTANCES CURRENTLY EMPLOYED BY BIG WINGS

POSITION INFORMATION

BRIEFLY STATE REASON FOR INTEREST IN A **PILOT** POSITION WITH BIG WINGS AIRLINES

ARE YOU WILLING TO RELOCATE?	GEOGRAPHIC AREA PREFERRED	CITY NEAREST TO YOUR HOME SERVED BY BIG WINGS

HOW SOON AFTER NOTIFICATION CAN YOU REPORT FOR WORK?	ARE YOU WILLING TO WORK ANY SHIFT INCLUDING NIGHTS, WEEKENDS, HOLIDAYS? ☐ YES ☐ NO

HAVE YOU EVER FILED AN APPLICATION WITH BIG WINGS? IF SO, WHEN? WHERE? POSITION?	WILL YOU ACCEPT TEMPORARY OR PART-TIME EMPLOYMENT?

HAVE YOU EVER BEEN INTERVIEWED BY BIG WINGS? IF SO, WHEN?	WHERE?	WHAT POSITION?

HAVE YOU EVER BEEN EMPLOYED BY BIG WINGS? WHEN? WHERE? POSITION?	WAS EMPLOYMENT ☐ PART - TIME ☐ FULL-TIME ☐ TEMPORARY

EDUCATION

ATTENDED FROM MO/YR	TO MO/YR	NAME AND ADDRESS OF SCHOOL	MAJOR COURSES STUDIED	DID YOU GRADUATE?	DEGREE RECEIVED	GPA
		LAST HIGH SCHOOL ATTENDED				
		COLLEGE OR UNIVERSITY				
		COLLEGE OR UNIVERSITY				
		OTHER SCHOOL (TECH, VOCATIONAL GRADUATE) AND ADDRESS				

LIST ANY SCHOLARSHIPS, ACADEMIC HONORS OR SPECIAL ACHIEVEMENTS:	EXTRACURRICULAR ACTIVITIES

Fig. 3-4. Sample application for "Big Wings Airlines." The application you fill out will be very similar—if not identical.

Fig. 3-4 (continued).

CERTIFICATES AND RATINGS

TYPE	NUMBER	DATE ISSUED	RATINGS, CLASS		
ATP					
ATP WRITTEN	N/A		GRADE		
COMMERCIAL					
INSTRUMENT					
FLIGHT ENGINEER	N/A				
FLIGHT ENGINEER WRITTEN			GRADE	GRADE	
			FEX	FEJ	FEB
MEDICAL CERTIFICATE	CLASS		LIMITATIONS ☐ YES ☐ NO		
			LIST:		
FLIGHT INSTRUCTOR					
GROUND INSTRUCTOR					
RADIO OPERATOR					

PILOT HOURS - ALL FLIGHT TIME MUST BE CERTIFIED

TYPE	NAME AIRCRAFT	AIRCRAFT COMMANDER/ CAPTAIN	SOLE MANIPULATOR/ PRIMARY/FIRST PILOT	FLIGHT INSTRUCTOR	CO-PILOT/ SECOND PILOT	DUAL RECEIVED	TOTAL	SIMULATOR
SE PISTON								
ME PISTON								
TURBOPROP								
SE JET								
ME JET								
HELICOPTER								
OTHERS								
TOTALS								
TOTAL PILOT TIME (DO NOT INCLUDE SIMULATOR OR 3RD PILOT HOURS)								N/A

OTHER FLIGHT TIME

TYPE FLYING	HOURS	TYPE AIRCRAFT
FLIGHT ENGINEER		
NAVIGATOR/RADAR OPERATOR		
3RD PILOT		
OTHER		
TOTAL NON-PILOT TIME		

INSTRUMENT TIME: ACTUAL _____   HOOD _____   SIMULATOR _____   TOTAL _____

ACCIDENTS _____
     (IF NONE, SO STATE)

VIOLATIONS _____
     (IF NONE, SO STATE)

REPORTABLE INCIDENTS _____
     (IF NONE, SO STATE)

MILITARY SERVICE

ENTRY (FROM) MO./YR.	RELEASE (TO) MO./YR.	MILITARY EDUCATION			MILITARY OCCUPATION ASSIGNMENT		
		NAME/TYPE/SCHOOL	COURSE TAKEN	NO./MOS.	MILITARY OCCUPATION	RANK	NO./MOS.
BRANCH							
RANK AT SEPARATION		ARE YOU A MEMBER OF THE ACTIVE RESERVE OR NATIONAL GUARD? ☐ YES ☐ NO			DATE COMMITMENT ENDS		
		HOW OFTEN DO YOU ATTEND MEETINGS?					

Fig. 3-4 (continued).

EMPLOYMENT HISTORY

Starting with your PRESENT or MOST RECENT EMPLOYER list in consecutive order ALL EMPLOYMENT and PERIODS OF UNEMPLOYMENT for at least the past TEN years. (Additional employment record sheet will be furnished upon request.)

PRESENT EMPLOYER	FULL NAME OF COMPANY			TELEPHONE (AREA CODE)	SALARY BEGIN	END
EMPLOYED	STREET ADDRESS	CITY	STATE	ZIP		
FROM — TO						
MO./YR. — MO./YR.	TITLE OF YOUR POSITION	NAME AND TITLE OF SUPERVISOR			REASON FOR LEAVING	
AIRCRAFT TYPE	CAPTAIN (HRS.)	CO-PILOT (HRS.)	OTHER DUTIES			

PREVIOUS EMPLOYER	FULL NAME OF COMPANY			TELEPHONE (AREA CODE)	SALARY BEGIN	END
EMPLOYED	STREET ADDRESS	CITY	STATE	ZIP		
FROM — TO						
MO./YR. — MO./YR.	TITLE OF YOUR POSITION	NAME AND TITLE OF SUPERVISOR			REASON FOR LEAVING	
AIRCRAFT TYPE	CAPTAIN (HRS.)	CO-PILOT (HRS.)	OTHER DUTIES			

PREVIOUS EMPLOYER	FULL NAME OF COMPANY			TELEPHONE (AREA CODE)	SALARY BEGIN	END
EMPLOYED	STREET ADDRESS	CITY	STATE	ZIP		
FROM — TO						
MO./YR. — MO./YR.	TITLE OF YOUR POSITION	NAME AND TITLE OF SUPERVISOR			REASON FOR LEAVING	
AIRCRAFT TYPE	CAPTAIN (HRS.)	CO-PILOT (HRS.)	OTHER DUTIES			

PREVIOUS EMPLOYER	FULL NAME OF COMPANY			TELEPHONE (AREA CODE)	SALARY BEGIN	END
EMPLOYED	STREET ADDRESS	CITY	STATE	ZIP		
FROM — TO						
MO./YR. — MO./YR.	TITLE OF YOUR POSITION	NAME AND TITLE OF SUPERVISOR			REASON FOR LEAVING	
AIRCRAFT TYPE	CAPTAIN (HRS.)	CO-PILOT (HRS.)	OTHER DUTIES			

PREVIOUS EMPLOYER	FULL NAME OF COMPANY			TELEPHONE (AREA CODE)	SALARY BEGIN	END
EMPLOYED	STREET ADDRESS	CITY	STATE	ZIP		
FROM — TO						
MO./YR. — MO./YR.	TITLE OF YOUR POSITION	NAME AND TITLE OF SUPERVISOR			REASON FOR LEAVING	
AIRCRAFT TYPE	CAPTAIN (HRS.)	CO-PILOT (HRS.)	OTHER DUTIES			

OTHER EMPLOYMENT: (IF THERE ARE ANY PERIODS OF UNEMPLOYMENT AND OR PART-TIME EMPLOYMENT NOT COVERED ABOVE, PLEASE EXPLAIN.)

Fig. 3-4 (continued).

SUPPLEMENTAL INFORMATION

WHAT FOREIGN LANGUAGES DO YOU SPEAK?	READ?	WRITE?

DO YOU HAVE A VALID DRIVER LICENSE? ☐ YES ☐ NO	STATE ISSUED	EXPIRATION DATE	LIST RESTRICTIONS. IF NONE, SO STATE.

HAVE YOU EVER BEEN DISCHARGED FOR MISCONDUCT OR UNSATISFACTORY SERVICE OR FORCED TO RESIGN FROM ANY POSITION? IF SO, GIVE DETAILS.

HAVE YOU BEEN CONVICTED OF, PLEADED GUILTY TO, FINED OR PLACED ON PROBATION FOR THE VIOLATION OF ANY LAW, GOVERNMENT REGULATION OR ORDINANCE? (EXISTENCE OF A CRIMINAL RECORD DOES NOT CONSTITUTE AN AUTOMATIC BAR TO EMPLOYMENT). IF SO, PROVIDE FULL EXPLANATION:

DO YOU HAVE ANY IMPAIRMENTS, PHYSICAL, MENTAL OR MEDICAL WHICH WOULD AFFECT YOUR ABILITY TO PERFORM THE JOB FOR WHICH YOU HAVE APPLIED? ☐ YES ☐ NO (IF YES, EXPLAIN)

HAVE YOU EVER SUFFERED ANY PHYSICAL INJURY? ☐ YES ☐ NO (IF YES, STATE WHEN AND THE NATURE OF INJURY?)

DO YOU HAVE ANY IMPAIRMENT OF SPEECH, HEARING OR SIGHT? ☐ YES ☐ NO (IF YES, PLEASE DESCRIBE.)

UNCORRECTED VISION

L _____ R _____

APPLICANT CERTIFICATION AND AGREEMENT

"I HEREBY AFFIRM that all of the information entered by me on this form and on any other forms completed at the time of my application for employment is true and correct. I understand and agree that any misrepresentation or concealment of information, regardless of when it is discovered, will be sufficient grounds for dismissal from or refusal of employment.

I HEREBY AUTHORIZE Big Wings Airlines, Inc., to request, and also authorize and request each former employer, educational institution and other persons or references listed, to furnish at any time, any information that may be sought concerning me or my work, habits, character or skill, or any other date required, whether in connection with my application for employment, for purposes of complying with surety company re-

quirements, or for completion of required background investigations.

I UNDERSTAND THAT I must be physically and mentally fit to perform the work for which I have applied. I agree to submit to medical examination(s) by physicians of the Company's selection and understand that if I fail to pass such examination(s), I will not be employed. I hereby authorize the Company's medical examiners to disclose to the Company any and all findings and conclusions arrived at during any such examination(s).

I FURTHER UNDERSTAND THAT if employed, I shall be required to sign a statement of employment conditions."

SIGNATURE _____ DATE _____

THANK YOU FOR COMPLETING THIS APPLICATION; IT WILL REMAIN UNDER CONSIDERATION FOR SIX MONTHS.

YOUR INTEREST IN BIG WINGS AIRLINES IS APPRECIATED.

AN EQUAL OPPORTUNITY EMPLOYER

Big Wings Airlines invites all Vietnam veterans, disabled veterans and other handicapped individuals to identify themselves. Identification is voluntary and reasonable accommodations will be made for such individuals.

Pilots receive good training at these airlines, but their financial strength is questionable. For such airlines, there are statements that are fine for application purposes. These companies know they aren't attracting cream-of-the-crop types. Generally, pilots applying at these airlines have only one purpose in mind—to build flight time and get on with a stronger carrier. So the idea here is to play the game. Address such items as your liking the domiciles offered by the airline and the fact that you realize that turnover at the airline is particularly high, which appeals to you because you may get to move up to Captain more quickly. The airlines *want* to believe a story like this because they are desperately seeking personnel who will stabilize their hiring situation. A good statement to make, then, might be: "I find the present level of hiring at Eastern to be very encouraging. It indicates that there is a position for me that will lead to a Captain's slot faster than any other airline."

Other questions that must be answered in the affirmative are the shift work and holidays questions. If you won't work weird shifts or holidays, don't even bother learning to fly. Also, the question "How soon can you report?" should be answered "Immediately." To come this far and not be willing to show up on any day they need you is very counterproductive to securing that fat job you want. You shouldn't care if it harelips the devil; if they say "Be here tomorrow," *be there!*

As far as the next headings on the application go—"Education, Certificates, and Flight Time"—answer these according to what is on record. All these items can—and most probably *will be*—verified. Some will be verified before an invitation to an interview will be offered.

Note the line for "Accidents and Violations" under the flight time heading. I hope you don't have anything to add here. But if you think there might be, call the FAA records office in Oklahoma City. If they have a record of the event in question, then note it on the application and do the best job possible to explain it in a favorable light. If the FAA record is clean, leave the answer line blank. The airline can know only what appears on the official FAA record.

Finally, under the heading "Supplemental Information," answer all questions thoroughly whenever there is a government record that can be checked. DWI (Driving While Intoxicated) records are being watched by the FAA. If you have such a record, report it to the FAA and the airline.

As far as reporting any injury you might have had at some point, don't be too ambitious to supply information that the airline will not be able to obtain. The airline just doesn't know who your doctor is, so they won't be able to check with him. If, say, you broke your foot in 1981, unless the doctor put it back on backwards and that causes you problems, don't supply the information. The idea is to appear healthy and non-accident prone.

Once the application is complete and you are ready to send it in, do yourself a favor: Make a photocopy of the completed application. This is for your protection and to use for preparation for the interview. As a general rule, make a copy of everything that is sent to the airline. The reason for this is so you can review the information that you have sent to each company. That way, anything the interviewer asks you will be on record and your reply will be in keeping with what is on the page in front of him. Nothing could be worse than answering the question, "Did you work at XYZ on such and such a date?" by telling the interviewer you "think it might be" another date. Your answers should be consistent with the information on file. Get it right the first time and stick to that story. Things like this go a long way towards making the interview routine.

The Contact File

No part of this whole process is as important as the contact file. Believe me, I've never met a pilot hired since 1970 who did not maintain some version of a contact file. That's just how necessary they are.

So what is a contact file? It is an alphabetized file where all pertinent information is arranged. By glancing at one airline's file card, you can immediately be reminded of their address, phone number, and—most importantly—the name of the contact person. Secondarily, the card should reflect what information is on file and when the last time a flight time update was sent.

Shown in Fig. 3-5 is a sample file card that would be filed alphabetically in some sort of card holder. A recipe-type file container is a good choice; all pertinent information can be written down on three-by-five-inch index cards. Another choice is a Rolodex-type system. These work well, but the cards are smaller and will not hold as much information. Use whichever you find more convenient.

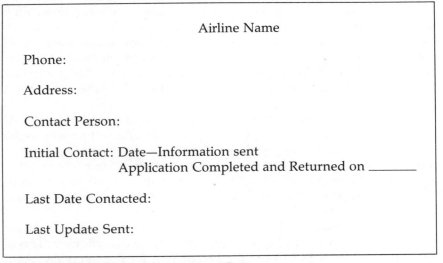

Airline Name

Phone:

Address:

Contact Person:

Initial Contact: Date—Information sent
Application Completed and Returned on _____

Last Date Contacted:

Last Update Sent:

Fig. 3-5. Sample index card for your contact file.

Let's go through the sample card so I can stress a few points. I put the phone number first because it is often the most important item—or, at least, the most used. The phone number might be the company's general offices or its personnel office number. The first number that you enter here might not be the number that you end up using. Finding a primary number to use for checking up on the status of your application can be difficult. It will probably come from someone you know in the aviation industry. Many airlines, for example, have key people who decide who will be called in for an interview. Such a person might be a chief pilot, a personnel director, the director of operations, or even someone's secretary.

Probably the single most important tip in this book is for you to *find out who this key person is*. At some airlines you will not be hired until you somehow make contact with this person. This is not to say that all airlines operate this way, but asking questions will identify those airlines that do, and possibly help you identify that all-important person.

Once you know who that contact person is, be sure that their office has a copy of your resume and application. I learned this the hard way. For years, unknowingly, I sent my applications and resumes to the North Central/Republic Airlines general offices to no avail. Quite by accident a copilot of mine at Metro Airlines, where I was employed at the time, mentioned that he had turned down a job at Republic for personal reasons. I had grown to believe I was too old, and what he said piqued my interest because he was my age. I began asking questions such as; "How did you get an interview? Who do you talk to?" Within ten days I was being interviewed; within 20 days I was hired. Years had gone by without my knowing the secret to getting hired at Republic! A few questions changed all that—and quickly!

Keep in mind, though, that a contact person is not to be abused! According to what they tell you over the phone, make up your mind when to call again. If they are in a hiring frenzy, call as often as every two weeks. Get your name known. On the other hand, if hiring is slow, one call a month might be enough—or even too much.

Always be courteous. A good way to start the conversation is like this: "My name is John Doe. I sent your office an application and resume about a month ago and because I hear you are hiring, I would like to check the status of my application." After they reply, you may inquire as to when they believe you will receive an interview. If they don't know, ask when you may check back with them, as you are *highly interested* in working for *their* airline.

The next three headings on the sample file card are self-explanatory. All dates of contacts should be entered. If you operate like I do, a review of each airline is necessary only about every other week. Generally, I did all my contact work on one day off. Then I coasted for a couple of weeks unless I heard something new through the grapevine. (Which reminds me—most rumors through the grapevine seem to be weeks old by the time you hear them. Therefore, when you hear an interesting rumor about hiring, react to it as soon as possible. You may already be too late.)

Finally, let's talk about updates. An update of your specifics should be sent to the airlines you are tracking any time there is something new to report. You just received your ATP, for example. Send an update.

Fig. 3-6. If you've done everything in this chapter, you're well on your way to your goal of a seat in the cockpit of an airliner such as this American DC-10.

Flight time, however, is a different matter. It takes about six months to accumulate enough new time to talk about, if you are working as a pilot. Therefore, every six months is often enough to send a flight time update. Nonetheless, send them regularly. A thick file folder is conspicuous—and anything that makes someone look at your folder is good.

So that's about it. In a nutshell, keep in touch with potential employers and find who that contact person is. Call the contact person regularly, but no so often as to become a pest. Then send regular updates that reflect movement in your career, such as an upgrade to Captain, a new license, or special training. Any excuse to add to your file and make them look up your folder is worthwhile. The exception to this is, of course, minimal increases of flight time. Once you are within the hiring parameters, a six-month update is often enough.

Follow this simple plan and be patient (Fig. 3-6). They will get back to you, and then it's up to you to impress them in the interview. That's the subject of the next chapter.

4

Your Airline Interview

IF YOU have the right qualifications, the system described in Chapter 3 will inevitably bring positive results. It will be all for naught, however, if you show up for the interview in a polka-dot suit. As with all the previous aspects of this program that we have addressed, the interview requires preparation. This preparation begins with anticipation of questions that will be asked, planning what you will wear, and learning whatever you can about the airplane that will be the basis for the simulator check ride. In this chapter, I'll discuss these areas, plus a few more, in the order that you'll have to deal with them.

Dress for Success

Several years ago, a best-selling book called *Dress for Success* was published. This book is an instructive guide for people who want to present themselves in a conservative and positive manner in the workplace or for job interviews. The book is a good place to start getting an idea how everyone else at your interview will be dressed—that is, the candidates, not the interviewers. Having been through many of these interviews, I can help you in this preparation. I assure you that I'm no expert on fashion, but I have been through the process and have seen those people who are successful and what they wear.

Conservative is the watchword when dressing for an airline interview. In fact, you must take it nearly to the point where you look like a clone of the guy or gal sitting next to you in the waiting room. This is unfortunate, especially if you are a somewhat flamboyant or stylish dresser. One thing you cannot be at an airline interview is a clothes horse. Possibly the *worst* thing you can do at an interview is stand out from the crowd because of the clothes you have chosen to wear (Fig. 4-1).

Fig. 4-1. You'll never get to sit "up front" in a DC-10 by being a "fashion leader." Conservatism is the watchword when dressing for your airline interview.

So what is the *right* thing to wear? Let's begin with the shirt. A conservative white is always acceptable and is by far the color most frequently chosen by interviewees. A very light blue or gray is also a good choice. Any color you decide on should be a solid color. Colors that I would discourage are in the pastel range. Where a light yellow might do, a pastel green will not, nor will a pink or pastel orange. In an airline interview environment, these colors might be perceived as "weak" or "effeminate" on men. Wear them to flight training or ground school with no fear of reprisals, but for the interview, go ultra-conservative.

Here's a little tip for choosing a dress shirt: consider your face shape. For example, my face is more of an oval with a hint of double-chin. I have a thick neck, like a running back. Experts on style advise picking shirts with long collar points when you have a rounded or oval face shape. Conversely, with long or square faces, a shirt with short or rounded collar points will tend to flatter the face more. I have found this to be true, so I use it when choosing my dress shirts. It might be good advice for you.

For women, the same color choices hold true. I must say, however, I have seen women candidates dressed only in white or gray. Usually these women's tops have been the types with collars that can be tied into a bow or one overhand knot.

Next, you must choose a suit. The color most often worn by airline applicants is navy blue, hence the clone look begins. The suit can have thin pinstriping, which is one way to set yourself apart from the next person without stepping over the line. Any navy blue fabrics that have a plaid look should not be considered. As before, this advice holds true for women, too. Wear a navy blue suit with a skirt. Pants on women generally send a combative (feminist) type of signal and should be avoided.

Women have been more successful with varying suit colors than men, and this could be due to the fact that we regularly see women in more colorful clothes. It might also relate to the fact that the airlines must hire just about as many women as they can to meet antidiscrimination requirements. Whatever the reason, there are more alternate choices available to women. Black suits with a nice pin or gray suits of a solid fabric color are quite acceptable. Plaids in gray might be acceptable, depending on other present-day styles. If gray plaids are popular in everyday wear, a woman will do fine wearing this style of fabric.

Men have only the alternate choice of gray as a suit color. Brown is *out*. So are sport coats. Simply expect to buy an acceptable suit, guys. *Don't* go to your interview in a plaid sport jacket that looks like a hand-me-down from Herb Tarlek on the old TV sitcom *"WKRP in Cincinnati."* You'll be shooting yourself in the foot. Solid grays are better choices for men, unless you can find a plaid that is almost invisible.

How about the choice between a two- or three-piece suit? In the 1970s, three-piece suits were much in vogue. In the late 1980s, three-piece suits can still be found but they have lost their place as the apparel of choice. The two-piece is perceived as a little more casual, yet still dressy, and it is chosen most often. Because I'm not going to try to predict what styles will be in vogue next, I will just advise you to watch Wall Street. Whatever those guys and gals are wearing on the job (except for their choice of ties), follow their lead and you will be in good shape.

The "conventional" advice on men's ties is something I have a little trouble swallowing. Since the book *Dress for Success* was published, most everyone has been wearing dark red ties as if there were no other colors available. I love color and prefer different tie colors (than red) simply because they *are* different. In the real world, however, red is the most often chosen basic tie color. Other colors such as dark blues with red accents are also good.

What about the yellow "power" tie? Even if Wall Street thinks that yellow is a power statement, stay away from it for an airline interview. The idea is to dress in a way that makes the statement: "*I am conservative.*" The airline interprets the "I'm conservative" statement as meaning "I am a safe pilot. I like to fly by the book. I like to fit in and therefore adhere sharply to cockpit standardization." It's crazy. All of us are individuals—but not on interview day!

What about footwear? Any dark shoe will do, but try to stay away from brown. Frank Zappa said it all in a 1968 song called "Brown Shoes Don't Make It." Brown shoes, although they are often worn with blue suits these days, are still not the most conservative choice, nor are they the best choice for an overall finished look. Black is fine. Dark navy blue is fine. And smoky gray shoes work well with a gray suit. Most airlines require plain-toe shoes for their uniforms, but for the interview, shoes such as the classic wingtip are quite acceptable. For the ladies, a low (not entirely flat) to medium heel is the norm. Flats don't have the overall class you want to project, and high heels tend to make a woman appear more on the sexy side than the business ("I can compete in a man's world") side.

As we all know, styles change. However, it is safe to assume that conservative business attire will remain much the same as it has been over the life of this book. I can't foresee fish ties and polka-dot suits becoming the norm any time in the near future. Just remember, look to the professional business community for hints on how to dress, and dress as conservatively as possible.

What about facial hair, mustaches in particular? Do you get rid of it if you have one? I did, but I think it really isn't necessary. Every airline accepts mustaches as normal nowadays. Don't sweat it. Don't show up with a beard, though.

Once you've picked out your clothes, it's on to the next step—preparing for the interview.

The Initial Screening Interview

Every airline's procedure is different. Some give a mini-physical the first time you visit, or even before you get to talk to the people responsible for hiring. However, the process is much the same for most airlines and many of the same questions are asked.

Airlines are different, however, so I must once again stress the importance of checking with FAPA. FAPA keeps a current counseling tape for every airline that is hiring. On that tape will be information on the exact sequence of the hiring process used by each airline. The tape will include any peculiarities particular to each airline, as well as their pet questions. It is impossible to present all that information here, but some practices are common to all airlines, and we can cover those practices here.

In every interview there will be one thing that sets you apart from all the lookalikes in the lobby. That is your personality. So rule one is: *Be yourself*. Unfortunately, some personalities do not fit in with some airlines. If a person happens to be that type—whatever "that type" *is*—then they very well may be turned down. If this happens to you, you will no doubt feel bad. The airline might have done you a favor, however, for you could wind up being employed at an airline that *does* like you, and end up happier to boot.

Many books have been written on how to perform during business interviews. All of the tips given in these books apply in this situation as well. But airline interviews differ because most interviewers check your logbook during the first part of the encounter. Therefore, you should prepare well ahead of time—even before making applications—to see that all columns add up. When airlines are able to be choosy—and who knows when that might be—they will use logbook inaccuracies to eliminate candidates.

The sort of inaccuracies I refer to didn't even occur to me until I had been through my first interview. To me, all the entries were accurate and true—no problem, right? Wrong! Due to careless attention, oversight, whatever, the columns at the bottom didn't add up across the page. Day column plus night column didn't add up to total time. PIC plus dual didn't add up to total time. Turbine time plus single time plus multiengine didn't add up to either of the other two answers. In general, the logbook was screwed up. It wasn't that I didn't fly those hours, but to the airline, *that is exactly what happened*. The result was an embarrassing point against me. If you are not ready with a properly tallying logbook, it could make you red-faced, too.

As I said, FAPA will have all the up-to-the-minute information that you'll need to prepare for a particular airline interview. Below follows a list of common questions that might be asked at some point during the entire hiring process at any airline. In other words, these questions could be asked at the initial interview, the Captains' Board, or even by the simulator check pilot. Consider each question, formulate what you believe to be an acceptable, concise answer, then practice that answer so that it can be delivered in a smooth, positive, impressive manner. I would also like to point out that although I was asked many of these same questions at interviews, most interviewers were most polite and did not try to put me on the spot. The best interviews for me and therefore (I think) for any applicant are where the answer to one question led naturally to the next question. In this type of situation, the interviewer becomes genuinely interested in the personality of the candidate (you, for example), and pursues a line of questioning that reflects that genuine interest. When this happens, it turns the interview into a conversation and not the traditional adversarial type of meeting.

- Why do you want to be an airline pilot?
- How did you become interested in flying, and when?
- When did you decide to become an airline pilot?
- How do you feel about becoming a flight engineer for several years?
- What are the keys to a safe airline operation?
- Where is our maintenance base?
- What do you know about our company?
- Do you know anyone who works for our airline?
- Can you afford to live on a probationary salary?
- Do you know what kind of salaries we offer?
- What type of aircraft do we fly?
- Where is our training carried out?
- Where are our major routes?
- What are the major concerns of the aviation community today?
- Do you own any of our stock?
- Do you know what price our stock is trading at?
- Where is our headquarters?
- What are the president's and chairman's names?
- Would you like to become a management pilot?
- What are your thoughts on deregulation?
- Where are our domiciles?
- Which domicile do you prefer?
- What will you do if we do not hire you at this time?
- What do you think about joining a union?
- What do you think about the seniority system?
- What do you think about the two-pilot cockpit?
- As a male, what would you think about copiloting for a female Captain?
- Do you always want to fly, or do you want a management position?
- Could you give a fair check ride to a friend?
- What other airlines have you applied to?
- What other airlines would you like to work for?
- Were you in the military? How did you like it?
- Tell us about your educational and flying background?

- How did you like school?
- What aircraft have you flown?
- Which did you enjoy the most/least? Why?
- What was your most memorable flight?
- What was your most unenjoyable flight?
- Have you ever had to divert to an alternate airport on a missed approach?
- Have you ever received an FAA violation?
- What do you think about private planes flying into major airports?
- What is the most dangerous thing about flying in general?
- What would spouse/friends say your best/worst qualities are?
- Have you ever been arrested?
- Have you ever had a weight problem?
- Have you ever taken illicit drugs?
- Do you drink alcoholic beverages?
- Are all the statements on your application the truth?
- Have you ever had your credit suspended?
- Have you ever been sued?
- Do you have a savings account?
- Do you own or rent your home?
- Is your family in favor of this job?
- Do you fix your own car?
- What are your hobbies?
- Why should we choose you for our company?
- Do you do any other flying besides your regular job?
- Would you accept overseas employment?
- Where did you grow up?
- In what percentage of your class did you graduate?
- What makes a good airline Captain?
- If you got two offers from two airlines at the same time, which would you choose?
- If you were flying good trips and the company called you in to be a full-time simulator instructor, what would you do?
- If you could have only three instruments in the cockpit, which three would you want?

The Stanine and Psychological Tests

Before you get hired, you'll most likely be given a battery of tests. Stanine tests, much like IQ tests, are designed to measure or quantitate your intelligence, personality, and psychological and aptitude profiles. The airline already knows you are crazy or you wouldn't be in this profession, so next they want to find out if you have the ability to do what is required of an airline pilot. That's certainly a reasonable attitude for a company that is going to spend a fortune on your training.

Stanine is a contraction for ''standard nine'' tests. It has to do with the way the tests are evaluated. In fact, all tests can be graded using the standard nine format. Below is an example of the scoring method:

Stanine Score	Portion of Sample		Cumulative Total
1	Lowest	4%	4%
2	Next	7%	11%
3	Next	12%	23%
4	Next	17%	40%
5	Next	20%	60%
6	Next	17%	77%
7	Next	12%	89%
8	Next	7%	96%
9	Highest	4%	100%

As you can see, each Stanine score represents a tenth for each bracket. Then some percentage of the applicants taking the test fall into each group of the Stanine bracket. For instance, if your score falls in number 5, half the applicants would be below you and half would be above your score. The goal is to be in bracket 8 or 9, and a little preparation will do a lot toward achieving that goal. If you want to practice, get one of the following books:

Airline Pilot Employment Test Guide by St. John. Published by Aviation Book Company, 1640 Victory Blvd., Glendale, CA 91202. Phone 213/240-1771.

Professional and Administrative Career Examination Guide. This is available through most large bookstore chains such as Waldenbooks and B. Dalton.

The Simulator Check Ride

The simulator check (which may or may not come in the order that I've used in this chapter) is used for two main purposes: to check your airmanship and instrument flying skills and to see how well you adapt to a strange airplane and to learning in a strange (and stressful) situation. Of course, if you have time and experience in the particular airplane type, the simulator check will be a more familiar event. The level of performance expected of you, however, will be higher than anything you've ever done before (Figs. 4-2 through 4-6).

Most of us have never flown a jet by the time we get to the interview process, but an airplane is an airplane and certain things apply to all of them. For a specific airplane there are, well, specifics. So, you are probably wondering, is there anything an applicant can do to prepare ahead of time for this event? Yes, there is, but first let me back up a bit.

There are two primary ways simulator checks are given. One is to give each candidate all the information needed for flying the program.

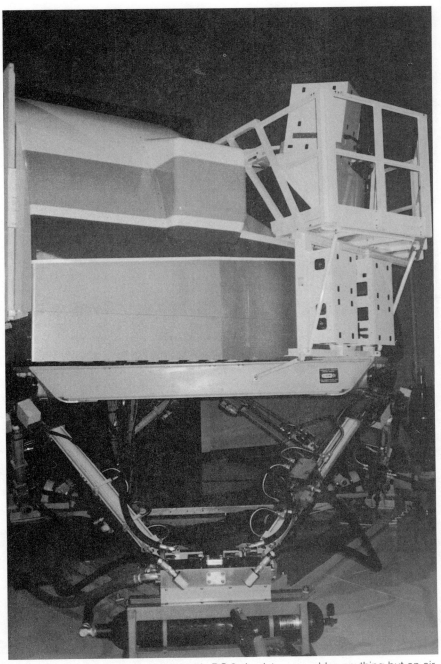

Fig. 4-2. Upon its eight chrome legs, this DC-9 simulator resembles anything but an airplane.

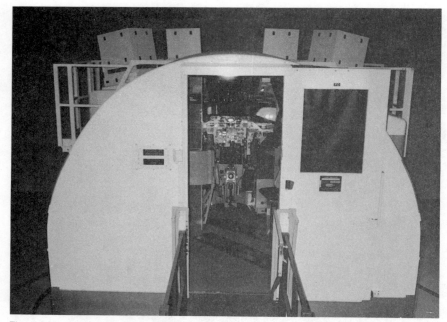

Fig. 4-3. A drawbridge leads to the door of the simulator and is retracted for "flight." Note the four CRTs mounted on the windshield of the simulator. These provide visual information for the pilots.

Fig. 4-4. The inside of a simulator (such as this Northwest DC-9) makes any pilot feel at home.

Fig. 4-5. The simulator that I took my sim check in looked like this one. Notice how the engine instruments are clustered just behind the thrust levers. The EPRs are on top followed by N_1, EGT, N_2, with the fuel flow indicators just outside of the two N_2 gauges.

Fig. 4-6. This is the instructor's station. From here, he causes the malfunctions that create terror in the minds of the pilots up front.

The other is to let the candidates fend for themselves. No matter which philosophy you encounter, a little preparation will show that you are a superior performer.

Let's assume that the airline has called you back for the simulator check. What do you need to know about the airplane being used for the simulator check beforehand? To begin with, you can expect one or all of the following maneuvers.

- Normal takeoff.
- Steep turns.
- Stalls.
- Single-engine approach.
- ILS approach.
- Go-around.

Most simulator checks concentrate on the basics (takeoff, steep turns, and approach), but your instrument skills are of utmost importance and under the tightest scrutiny. Knowing a few facts about the airplane will make the check go much easier.

Let's take each category above one at a time and assume you have never flown a jet airplane before. If you have, there are still a few points here you might need to remember. These descriptions will help you think in terms of preparing for the check ride.

Normal Takeoff: This part of the program is like any takeoff. For jets, though, you will need to know that EPR (engine pressure ratio) is the main unit of power. Normally EPR is the top gauge on the engine instrument cluster, and takeoff power is marked by an adjustable "bug." The normal takeoff range for a jet airplane is between 1.85 and 1.95.

During takeoff, remember that any warning light or malfunction prior to the calling of "V_1" calls for an aborted takeoff. If this should be tried on you without discussion of procedures, just close the thrust levers and call "Abort." This will show your judgment and knowledge of basic procedures. Don't worry about spoilers or reversers. Unless you were told specifically to use them, it is not necessary to use them to show your competence in getting the airplane stopped. Besides, it is normally the Captain's job to pull the spoilers up or actuate the reversers, and you are being tested as a First Officer.

If you don't get any warning bells or lights or whistles during take-off, rotate smoothly, call "Gear up" after a positive rate of climb is established, and have the check pilot clean up the flaps for you.

Steep Turns: This is a test of your instrument scan and handling techniques. A jet handles like any other airplane with the exception that speeds will build up faster than propeller-driven airplanes if you let the nose drop. Flying in this phase will be in a clean, no-flaps configuration. What you need to know is at what speed this maneuver should be flown. This will vary from airplane to airplane, so find a pilot for the airline that will be testing you and ask what speed is used. (I know this information may be given during the check ride, but some airlines prefer to let the candidates sink or swim on their own. Investigate!)

Next, be aware that in flight, N_1 is most often used as the power indication instead of EPR. For example, 70 percent N_1 is often the normal power setting for steep turns, configuring for the approach, and flying the ILS approach. So remember, 70 percent N_1 is for maneuvering. This is a good starting point for any jet and is better than nothing in a situation where you have no other information to go by.

Stalls: It is doubtful that you will have to perform these for a simulator evaluation. If you do, remember this: Call for max power and level the nose to the donut on the attitude indicator. Wait for the airplane to accelerate out of the stall with the *nose level*. Do not drop the nose, as altitude loss will be very great in a slick jet. A normal reaction of a prop pilot would be to lower the nose below the horizon to pick up airspeed. *Do not* do this in a jet. Airlines have an aversion to altitude loss in stall recoveries. Enough said.

Single-Engine Approach: This procedure is rare in simulator check rides. (I have been told, however, that United Airlines does pull it on candidates, usually through the vehicle of an engine fire and subsequent engine shutdown.) In any event, call for the checklist by name. If there's a fire, then call for the engine fire checklist. If the engine flames out, call for an engine flameout checklist. In the event of a fire, you might suggest that the thrust level could be retarded to check for a bleed air leak, but only after the initial checklist has been called for. This would indicate some knowledge of turbine engines.

Other than that, single-engine or engine-out procedures are flown close to normal. Addition of some extra speed is called for in the event of a go-around. If no one's giving you the correct speeds to fly, use 10 knots over the book V_{ref} value for the approach. Usually there will be a speed book displayed with values for different flap settings somewhere on the panel, usually over the radar.

ILS Approach: No matter what, the candidate *will* see at least one ILS approach. By this time in your career, you'll have hundreds if not thousands of these behind you. The procedures in a jet are the same as any other type of airplane: Configure prior to the outer marker and remain stabilized down the approach glideslope to the flare. As I mentioned earlier, 70 percent N_1 is a good power setting. Reduce power to 70 percent and let the airplane slow down. If flap speed information has not been given, then use the speed book. (Again, the speed book is a little spiral notebook, often in some bright color, that has large numbers depicting flap settings and the corresponding slowest safe speed beside each setting.) In other words, the speeds displayed are minimum speeds that can be flown in any configuration. As a rule of thumb, add 20 knots to each value to know when to call for that increment of flaps to be extended. For example, if the book says Flaps 5 degrees—207 knots, call for "Flaps five" at 227 knots. This is an example and doesn't reflect any particular airplane.

Most intermediate flap settings run through 15 degrees prior to gear extension. After reaching 15 degrees of flaps, ask for gear down *and* the before-landing checklist. Once the gear is down, continue to extend flaps to the final increment, usually 30 or 50 degrees. Then, maintaining close to 70 percent power and bug speed (V_{ref}) plus five or ten knots, fly the airplane to the ground. Don't worry about the landing; your evaluation will not hinge on it. If your landing happens to be good, remember that "blind dumb luck will win out over skill and cunning every time!"

Go-Around: Most jet airplanes use 15 degrees flaps at moderate weights as a standard takeoff flap setting. This will also be the go-around setting. If you're told to go around, this is how most airlines will expect that maneuver to be executed. Call for "Max power, flaps fifteen." Rotate the nose to takeoff attitude, usually 15 to 20 degrees nose-up. (If in doubt, use 15 degrees.) Watch for a positive rate of climb, then call for "Gear up." Climb to missed approach altitude and clean up the remaining flaps. Also, remember 70 percent N_1 will keep you from accelerating through flaps limitation speeds when you level off. Practice these calls before the simulator check ride in case you need them.

Now, believe me, these tips are just that—*tips* in case you are given very little preparation information. The information here is very common to most airlines, and chances are you'll be given a full briefing. Still, tips such as 70 percent N_1 might not be mentioned. It's probably the most important single item that makes it possible for the airplane to be flown properly at slow speeds.

There is one more tip for maneuvering a jet airplane at slow speeds. I'll add it here so that you can consider it separately. Fuel flow is often used for setting power during the approach and descent phases of flight. It works this way: During vectoring for an approach, it is common to see fuel flows in the neighborhood of 2,500 to 2,100 pounds per hour (pph) per engine with power applied at about 70 percent N_1. For example, say you're on downwind at 6,000 feet and the controller assigns you a speed of 210 knots. Experienced pilots often set this speed by observing the fuel flow. Roughly 2,100 pph per engine will equate to 210 knots. Similarly, 2,500 pph will correspond to about 250 knots. In other words, by dropping the last digit, you can see the speed that will result from a particular fuel flow setting.

One last tip on simulator check rides: There are many companies now that offer simulator check ride preparation using sophisticated modern nonmotion simulators (more properly known as *training devices*). These companies can usually configure the training device to fly nearly like the airplane in which you'll be taking your check ride, plus these companies also keep on hand profiles of the check rides given by various airlines. (The way they obtain this information is easy, of course.

The companies simply ask their former students what occurred during their check rides, thus helping prepare the next round of candidates.)

The Captain's Board

You've made it this far. Few candidates get this far and *don't* get the job. Be that as it may, the Captain's Board is just another opportunity to flunk out. So what should you expect?

Walking into the room, you are suddenly face-to-face with two to six gentlemen and probably a lady all dressed in business suits. They are there to find out about your personal career. How did you get this far in your career? That's the question upmost in their minds. Next, are you likable and do you seem level-headed? Believe it or not, they want to see each candidate make it. They figure most of the menial selection processes have been completed and all they want to know now is: Will you fit in?

Most candidates look forward to this with dread, only to find that it is a pleasant experience. (It seems this feeling of dread never goes away in the airline business. Every time I fly with a Captain I've never met before, that same feeling comes over me. Will he like me? Will I like him?) Ninety-nine percent of the time it turns out to be a pleasant experience—more of a hangar-flying session where you get to talk and reminisce. Just don't talk too much about bad experiences.

You might again hear some of the questions listed a few pages back. Be ready with pat answers that don't *sound* pat. This technique will keep you in control of the interview and keep you out of hot water. You know what I mean? The tendency to stick your foot in your mouth is always ready to rear its ugly head. The only advice I have is go in to the Captain's Board looking to have a pleasant experience and *you will*.

The Company Physical

As I mentioned earlier, some companies begin the interview process with a mini-physical. There is little you can do to prepare for a physical at the last minute, so the key is to start early. It takes about two to three weeks, for example, to lower the blood cholesterol level (LD) by controlling the diet. Blood work will be done during this physical and the three main things the lab will report on will be cholesterol, triglycerides, and glucose tolerance (diabetes test). The first two may be lowered appreciably by cutting fats and sugars out of the diet or restricting them as much as possible. I'm sure you already know this, but for that airline job, take that extra step and limit your intake of fats and sugars whenever you can.

To draw upon my own experience, let me explain what I did. I dropped almost all ingestion of red meat, because it contains fat and cholesterol. Then I increased the amount of salads and vegetables I was eating. I'm not a big salad fan, but I do like vegetables. Next, I cut down on milk because of fats and additional calcium that can be construed as a possible arterial clogging agent. I cut out soft drinks that contain sugar and drank mostly water because it is a good cleansing and purifying agent through the kidneys. I ate a lot of fish and chicken and tried to have it prepared by broiling whenever possible to avoid the possibility of it being cooked in oil that contains saturated fat.

Basically, that's it in a nutshell. I lost a little weight and lowered those blood levels that might have been marginal. I became more healthy and passed the physical, too.

Follow this advice and you will pass. You'll be hired, and then it's on to the next phase of becoming an airline pilot—probation. We'll discuss it fully in the next chapter.

5

Your Probationary Year

IT MAY begin with a telephone call or a letter that reads like the one in Fig. 5-1.

As you read this letter your heart soars, for at last the coveted job of airline pilot is yours! Not to sound ominous, but it will be yours only *after* you endure probation. So let's talk about probation. Not much, if anything, has been written on the subject, yet it is the last and most important step to securing that seniority number at an airline.

First, let's discuss what probation is and then take an imaginary trip through the year of probation. This will give you a taste of what to expect as you enter the major airline lifestyle and give you a leg up on those who will go through the experience with you.

What Is Probation?

I imagine probation for airline pilots was thought up somewhere in medieval times, long before airplanes even existed. Some sadistic prince of some kingdom long forgotten probably issued this edict: ''We shall indenture our servants until such a time that they have provided the court with service that is equal to trust we put into them. As well, their compensation shall only be a subsistence to test their initiative, genius, and resources. Thereby shall they prove themselves worthy to wear the clothes of the kingdom.''

And that, my friends, is just about the story of probation. In plain terms, probation is seen by the airline as a one-year period during which the company closely examines your performance as a pilot and employee. If either of these categories are judged unacceptable, you are out of a job. Also, the company deems it necessary to pay a low starting salary due to the initial investment in training that they have already had to spend on each new pilot. Airlines use this time for evaluation because training costs can equal some pilots' salaries over the years. The airline must assure itself it has made a good choice in allowing you to

Miss Jane Doe
555 Four Fives Lane
Bountiful, Utah 80555

Dear Miss Doe:

You have been selected as a successful candidate to join Big Wings Airline as a flight officer. Your class date has been set as January 13, 1989. Further arrangements can be made by contacting this office at (555) 555-5555 before 5 pm during any business day. We congratulate you and wish you the best as you begin your career at Big Wings.

Best Wishes,

Dody Doolittle

Fig. 5-1. Congratulations! When you get a letter like this, you've made it—almost!

work for it. This flies in the face of what most of us believe, which is that we're *more* than qualified to fill the airline's pilot positions. As a result, you must adjust your attitude and be somewhat subservient and put up with all that goes on from the first minute of the first day of class.

The First Day of Class

You catch the van from the motel to the training center. It soon fills up with faces that may range from unfamiliar to very familiar. Everyone is dressed in ''clone suits'' and it looks and feels like another interview. Everyone is fighting anxiety. And everyone is proud that he or she is on this van for the beginning of what each believes to be the ultimate career.

Some of the pilots begin talking. Maybe they know each other from previous jobs. Questions start flying: ''When did they notify you? Where have you been flying? Do you know so-and-so? He told me so-and-so about Big Wings. How many airplanes did you hear they were buying? How many pilots are they going to hire?''

The van pulls to a stop in front of the training center. The driver calls, ''This is it, and good luck to you!'' Everyone nervously giggles and walks nearly silently to the front door steps (except, of course, for those two or three folks who really know each other well; they just keep talking).

Immediately the group finds its way to the classroom. There, on rows of tables, are company manuals, insurance forms, and possibly an airplane manual—and, most important of all, a large hand-printed name card *with your name on it!* It's not a dream after all! They are expecting *you* this day and have prepared for you ahead of time! A wave of comfort sweeps over you. You sigh with relief. This is real!

Finally, everyone has filed in and sat at their various designated places. You look around at their faces, wondering what is behind each one. All those faces are new acquaintances, people you will know for the rest of your life. Some will become close friends for life. The class includes members who have come from poor families and have struggled all the way to this pinnacle. Others have fathers who fly, or even own their own airplanes. They will claim to have the most information and the "inside track." What will soon be apparent is that no one at the airline has the most information or the "inside track." Performance in the position of pilot is all that matters. And rumors—they will be numerous. Your information and grasp of "the big picture" will come from fitting all the various rumors into a plausible plot—a feat that is only *somewhat* reliable after many years in the business.

Now the guys and gals begin to pick up the conversations that were started in the van. And if there is not already a list before you with seniority numbers, *those* questions will inevitably follow. Long before the instructor, vice-president, or whoever arrives for the welcoming speech, the class will have established who is senior.

The next day or so will be full of papers to sign and speeches to absorb. Union people will come in and lobby for union involvement. Training directors will come to paint a rosy picture of how many pilots will be hired following your class. *Beware!* This may or may not be accurate.

At last, the seniority list is presented and your position is cast in stone. Those names appearing above yours, with a lower number, will stay there. And those names below yours, with a higher number, will—thankfully!—also stay there (commonly referred to as the jerks above you and the scum below). So to someone, you'll always be scum. That's the way it is laughingly played and referred to in the airline world. Yet seniority is everything in your future.

Although it might seem that seniority numbers are never changed, let me tell you about a case that affected me and 90 other pilots in my seniority bracket. Ten days after I was hired at Republic Airlines, the merger with Northwest was announced. The final class for Republic was hired six months later. There were 194 pilots junior to me. Then the announcement of furlough came for the Republic pilots, while Northwest continued to hire new pilots. This was unacceptable, according to ALPA (Airline Pilots Association), the pilots' union—furloughing one group of pilots while hiring a new group. The solution was that furloughed Republic pilots were transferred to the Northwest side of the airline, and a total of 150 pilots were transferred, missing me by 44 numbers.

This arrangement was thought to be temporary while the seniority lists were being merged. As I write this, however, I pass my second anniversary with Republic—now Northwest—and still the seniority list is not merged.

What happened to those 150 pilots who were transferred? They are officially Northwest pilots and have advanced up the Northwest ladder already, with 29 of them landing right seats in the Boeing 757 and a few getting Flight Engineer slots on the "whale" or 747. Each one of these lucky pilots is making $30,000 to $50,000 per year *more* than each of the 91 Republic pilots left behind to fill the lowest-paying positions as Flight Engineers or Convair 580 First Officers. By virtue of being senior, I lost at least $30,000 during my second year with the airline. Being hired first can blow up in your face, and in these days of deregulation, anything is possible. Forewarned is forearmed.

Back to the subject: Eventually all the rhetoric passes and the hardcore lecturing on company procedures and airplane equipment begins. It is all interesting, and the new-hires take to the task of learning like a sponge to water. College was never like this—this is *big-time* education! A multimillion-dollar career is on the line here and no one, but no one, wants to blow it. The hours at night in the hotel get later and later. The lights go out only in fear that a new-hire won't get enough sleep to understand the next day's important subjects.

At last the first of several hurdles arrives: written tests. Passing is important. It is everything. Failing? It's unthinkable. No one fails. Everyone is pulling for the others in class. If anyone is having problems learning the material, classmates pitch in and tutor the slow learner. If anyone should fail, the bond of friendship within each class would suffer great shock. It's truly one for all, all for one—a brotherhood.

The importance of these initial written tests can best be measured by the feeling of accomplishment each pilot receives as he or she steps toward the goal of becoming qualified on an airplane. In other words, passing grades are more meaningful for each pilot in terms of making him comfortable about his progress than as a gauge for the airline to measure aptitude.

The next great challenge looms in the dim light of what could best be described as a huge garage. Perched high on eight or more hydraulic legs sits a box bearing little resemblance to a sleek jet airplane. In fact, it looks more like a Trojan horse type of mechanical war machine from a science fiction movie.

Inside this contraption airline pilots are made. They walk in cool, calm, and collected, and they exit tired, spent, and sweaty. The experience is real, and the challenge is great.

As a pilot for Metro Airlines, I had spent the last four years as a Convair 580 Captain. A Convair has no control-augmentation system (which uses hydraulic power to make controlling the airplane take less physical effort). Hence a great deal of physical effort is needed to control the Convair. DC-9s, on the other hand, can be flown with a much lighter touch. Therein lay my problem. There I was, a pilot used to applying large (20 to 40 pound) control pressures to make the airplane respond now facing a machine that used about a tenth of the effort to fly.

My best efforts to fly the DC-9 simulator looked ham-handed. After a few nights it began to look as if I was not airline material. My ILS approaches were anything but the precision approaches they were supposed to be. We rumbled down the electronic funnel, bouncing off one side, then the other. It was clear to the instructor, my partner, and myself that I had a massive overcontrol problem. Time was running out. We had the weekend off, then three more sessions. The last of the three would be my check ride.

I left for home that weekend shaken, wondering—because I had never before failed a check ride—would this be the one? Would I be out of a job? It was time to fall back on basics that had worked for me all my career. And so I did.

"Flying the bed," that's what I call it. I lay down on the bed with a checklist in my hand and began a flight from the first moments of pre-flight. With checklist in hand, I put myself mentally through all the routines right up to takeoff. My mind became focused on the job at hand. To me it became every bit as real as the simulator. Finally, I was airborne, performing all the check ride maneuvers as if they were actually happening. I could feel the controls in my hand. I caught things that I didn't have time to contemplate in the actual simulator environment and thus I could move those things to where I could best manage them within the confines of standard operating procedure. The result? My skills were sharpened, especially after several sessions of "flying the bed." This technique has never failed me and I expect to be using it again throughout my career.

As you may have gathered, I came back after that weekend to challenge the "box" again. My performance was much enhanced—so much so that the check ride was uneventful. This is not to say I flew perfectly, but I did fly competently and safely. Later, after some time in the airplane, I began to develop some finesse in flying it.

Normally, everyone will pass the simulator check ride and be on the line in short order. Once again, no one thinks of what might happen if they should fail. Everyone always thinks of passing.

The airplane checkout comes after the simulator check, and this part is usually a piece of cake. This is because the check airman cannot fail very many things in a real airplane. The gear is going to come down unless there is a *real* (uh-oh!) problem. And single-engine procedures? By now everyone has seen those until engine-out emergencies seem normal. The upshot of this is that the simulator is the big deal. The airplane check is a *little* deal.

Still, you'll walk away from the airport that night (most airplane checks are at night, when the airplanes aren't being used to make money) with a feeling of accomplishment. It is a happy time. Ground school and that ever-pressing need to study more will be gone. The pressure of check rides is behind you. For the present, the pressure lifts and your spirits soar. You feel finally that "I can do this job. I've proven it. I am a part of this operation now!'' You'll never forget that feeling, for it was a long time coming.

The next item on the agenda is IOE, or ''initial operating experience.'' For me it was the most unforgettable experience of all. The word *initial* sums it up—the first time you get to operate the airplane or panel if you're a Flight Engineer in the ''line'' environment. This is where a new pilot begins to learn the tricks of the trade. Even if the airline's training curriculum includes a LOFT (line-oriented flight training) program, it takes real experience on the line to learn the little things.

Two things stand out most about my IOE on Republic's DC-9. Although I probably remember more little things about that trip than any other trip in my entire career, skimming the rooftops on the hill overlooking the runway at San Diego was a most exciting time for me. It required everything I had learned about flying the DC-9 in the previous three weeks of flight school. The approach required precise speed control, precise positioning on the VASI, and precise touchdown with reverse thrust because San Diego's runway is not overly long—especially for a rookie! I remember that moment as a challenge and a joy.

The other moment I remember from that trip was not a victorious one, yet it taught me a lot about flying the DC-9-50. (The Dash 50, Fig. 5-2, is the longest version of the generic DC-9. The much newer MD-80 is longer.) First, I learned that there is an abrupt visual change in the picture out the window when one is about to botch a landing. I learned to recognize that change. I also learned never, but *never*, pull the power off in a Dash 50. It is an invitation to smack the runway, mightily! And we did—maybe not nearly as hard as the Eastern Airline pilots whose DC-9-30 cracked open on landing just behind the wing, but hard! Now I'm proud to say I never have landed that hard again. I learned quickly. It was just too painful to repeat.

Fig. 5-2. The transition paint job from Republic to NWA. This DC-9-50 sports part Republic and part NWA red tail.

And so it will go—IOE, then the absolute *real world*, with no check pilots looking over your shoulder, only a Captain who expects you to be up to speed with most of the other pilots in the company. So you walk down the jetway and onto the airplane. You look for a place to store your bags and make a nest for the flight. You haven't established a routine yet, so you fidget and try things until something clicks. From then on, that's how you'll do it. A pattern has been established.

Outside, you begin the walkaround, paying close attention to items that have no bearing on safety such as chipped paint or stains from the blue chemical in the toilets. There is a reason for this process. Obviously, you want to be safe and do a good job because probation is the game that you're playing. The most significant reason, though, is to assimilate what is *normal* wear and tear on the outside of the airplane and what is *not*. This information will serve you well over your career and it takes hundreds of walkarounds before you can absolutely know what is right and what is wrong on an airplane such as a DC-9.

Entering the cockpit again, you notice that the Captain has arrived and is already into his cockpit checks. Questions run through your mind: "Is this guy as mean as he looks? Is he a stickler? Does he hate to fly with new guys?"

He turns around and sees you. Instantly he shoves out his hand, states his name, and gives you a big smile. "Hey! This guy is all right," you surmise. And most of them are. That's almost always the way it is. They look grim, those Captains—so many of them with gray hair and stern looks on their faces. Yet almost always they enjoy the company of a fellow pilot, even a new one. Very few Captains will make you uncomfortable or make it difficult for you to do your job.

It's a good thing, too, because you will have to ask some of these Captains to evaluate your performance on a trip. We were supposed to ask Captains to fill out an evaluation prior to a trip, but most said just to remind them at the end to fill it out. That worked well for the guys in my class. If the trip went a little rough, well, we just sort of conveniently "forgot" about the evaluation. As a result, all the reports were favorable—not that they wouldn't have been anyway.

Probably the easiest way to please Captains during the probationary year is to fly absolutely standard procedures. In short, don't improvise. For example, standard procedure at Northwest is to lower the flaps to takeoff setting *after* the airplane has powered away from the gate and as it has rolled forward 25 to 100 feet. This assures that the flaps won't be lowered onto a baggage cart or some other obstruction. Thus, putting the flaps down too early might damage the airplane, upset the Captain, and ruin your career. This example is just a small one and is just the sort of thing that can label a pilot as a maverick. It's all right to be different *after* your first year.

Occasionally you will find a Captain with a peculiarity. If it doesn't conflict with company procedure for flying the airplane, disregard it and perform the function as the Captain requests. For example, the Captain might say, "I never land the airplane without my blue golf glove on." Several flights later, you realize the airport is getting close and the Captain hasn't yet put on his golf glove. The answer? Say, "Captain, don't forget that glove. I'd hate to see a string of such nice landings go down the tube." Instantly you have made a friend for life. And that evaluation he's supposed to fill out? It'll be one of your best.

That's how the game is played and anyone who rebels will find the going much harder, maybe even impossible. Don't get me wrong. You don't have to lick Captains' shoes. Just appease them. Appeal to their egos and they will respond to yours with compliments as well.

Finally, at year's end comes the final simulator check and evaluation. This time the airplane is familiar and you are a seasoned pro. Yet you must prove it this time as you will for years to come. The check pilot knows this is your probation ride and he wants to see you make it. Your simulator partner is a Captain who's just taking his 25th or so six-month check. He's relaxed and he also wants to see you shine through this ride. And you know what? You do. It all seems like old hat. The ride is a snap, but it freshens your outlook, attitude, and skills. You're a professional airline pilot and in the groove. Good people, good pay, and good flyin'.

There's just one more little hurdle and probation is history. You're assigned to a management pilot for an interview. He's the guy who puts the stamp of approval on your year. Sitting across the desk from him in his office, you watch as he thumbs through all those Captains' evaluations. You know they're good. And now so does he. He's got one or two questions. Most importantly, after a year, how do you like flying for Big Wings Airlines?

Don't blow it now! Be as enthusiastic as you were at your first interview. "It's better than I ever imagined," you tell him. "Fine," he says, "we're glad to have you for life." He offers his hand and you depart.

Outside the door you shed your second-class pilot stigma. It's reminiscent of the feeling you had the night you checked out in the airplane. You're one of the group, for good. When you make Captain, remember how it was, for the new pilot's sake.

A Savings Account

If there was only one piece of advice I could give a newcomer to this game, it would be to have a savings account. The probationary year can be a test of wits in personal finance. Although starting wages are three times higher than when your Captain started flying, your expenses can be *four* times higher. The only way to get through that year and not have outside financial distractions is to have something to fall back on. Some pilots fly for the National Guard and that serves the same purpose. The idea is that when the car breaks down or the baby decides to come before the probationary year's end, there will be money to cushion the blow to the monthly paycheck.

I'm not sure how much is enough. It depends on a person's lifestyle, but the more the better. I never have been a financial expert so I won't advise you on how much. Just see that you have something because every pilot in my class, as far as I can tell, needed that extra cushion and was glad he had it.

That was a quick glance into what the first year is like at the major airlines. Commuter or regional airlines are much the same, and you may be stopping off at one of them prior to coming to the majors. We've hit all the high points and I am sure your career will reflect many parallels. It's a different lifestyle and maybe the only one that will ever suit you. All I can do is advise you, and I have done that.

In the next chapter, we'll look at the future and see what may be in store for *your* career.

6

The Outlook for
Future Hiring

JUST THE other night, as I was riding the van to the motel, the driver, a young man, asked about his chances of becoming an airline pilot. "I've got my private pilot's license, what should I do now? How can I build time? What's the least amount of total time it would take to get hired by a regional airline? A major?" he asked. The other pilot in the van with me touched on most of the major points in this book while I listened. When the other pilot finished, I told the young man that I was writing this book and that I had all the latest information with me.

"Drop by my room," I said. "You can scan all the material and photo-copy what you like."

"Okay, after my next trip to the airport, I'll be up."

He never showed up.

Maybe he got the wrong idea, but here's the point I'm trying to make: First, *any* help *anyone* in this business offers you, *take it*. They might have just the key that will open the door of your career. And if they don't, they might have the key that opens a later door. That's called networking, and you build your network as I discussed in Chapter 3.

Enlarging on that point, I'd like to say that personally I want to and *will* bend over backwards for anyone who asks for my help. Numerous people have gone out of their way just for me, and I owe the consideration to others. Therefore, the young van driver, just by asserting himself, could have gained valuable up-to-date information about an airline career. He also would know one pilot at a major airline, one who has inside tracks to the nation's number one commuter airline. I'm not tooting my horn. You, the reader, never know who the person you're talking to knows. You must be assertive. It will save you hours of flight time, months in an unproductive job, and years in getting hired. It's just that important!

So what did we tell the driver about hiring? We told him it was good and getting better. In mid-1982 the recession ended and the first large-scale retirements and expansions began. The hiring soared. "All well and good," you say, "but how about the future?" Growth outstrips attrition and attrition alone is enough to precipitate lots of hiring. By 1997 we're expecting hiring to total 32,000 *minimum*. FAPA estimates the industry wide needs may reach from 42,000 to 52,000 pilots to fill all vacancies from airline entry levels, such as regional carriers, to nationals and majors. If that's not encouraging to you, then I don't know what is.

Perhaps you are facing a handicap, such as glasses, no college, or a weight problem. Hang in there. The airlines need pilots.

Visual acuity is becoming less of a barrier. I suspect that in the next three years, even the majors will be accepting 20/200. Face it: An airplane with no crew makes no money. In fact, it costs money for the airplane to sit. It may come down to: "If you're interested, we'll take you. No license? We don't care. Want to fly for our competitor? Take a hike."

People sitting at zero flight time and experience and on will have a 90 percent chance of making it to the majors by 1995. So get going! The barriers are falling—and in many cases, have already fallen.

Salary Range

ALPA has been negotiating several contracts with major airlines, and B-scales are eventually expected to be limited or even disappear, depending on the pilot shortage. Continental is having trouble attracting candidates because of pay, benefits, and the airline's style of management. Eventually Continental will have to reckon with this stance and alter its course. This means more benefits, higher salaries, or the company's pilots will simply go fly somewhere else. This in turn justifies other airlines paying more to attract and keep the best candidates, and so an upward escalation in salaries appears. When Continental calls 24 candidates for a class starting date and only half show up, the time is fast approaching when the airline will have to change its policies or park airplanes and go bankrupt again.

Finale

Folks, the dream is yours. It's a great job, but it takes training. Sacrifice is a word common to most airline pilots. With that sacrifice come prestige and financial reward. Regardless of what the news media have said about pilots in the last few years, the public still holds the position in the highest esteem. They must do this to have faith in the individuals who carry them higher and faster than ever before (Fig. 6-1).

And salary . . . well, a TV commercial I saw the other day puts things in perspective. It was for an insurance company. Here's a fellow who needs a financial plan, the ad claims. A yuppie type. Successful. He

Fig. 6-1. Your future? This artist's conception shows the joint U.S. government and Northwest Airlines hypersonic aircraft. Dubbed the "Orient Express," it's projected to fly from New York to Tokyo in just over three hours.

makes almost $30,000 a year. He's reluctant to admit he makes so much. Folks, that's only second-year pay at the worst-paying airlines. Enough said?

So keep at it. Write me in care of this publisher and let me know how you're doing, and if you need a little extra help, let me know.

Lots of luck to you and enjoy your new career.

Appendix

Colleges and Universities with Aviation Programs

The following is a partial selection of schools and colleges which offer flight training, and in many cases also a college degree. Check with the Federal Aviation Administration for a current list of certificated pilot schools.

Alabama

Alabama Aviation and Technical College
Office of Admissions
PO Box 1088
230 Hwy South
Ozark, AL 36360
Program: Flight training ratings: private through ATP. Also offers helicopter. Also offers two year associate degrees: General Aviation Technology

Auburn University
Office of Admissions
Mary Martin Hall
Auburn, AL 36830
Program: Flight training ratings: private through ATP. Also offers bachelors degree in professional pilot.

Gadsden State Jr. College
Office of Admissions
George Wallace Drive
Gadsden, AL 35999
Program: Private, commercial, instrument

University of South Alabama
Office of Admissions
308 University Blvd.
Mobile, AL 36608
Program: Private, commercial, instrument, multiengine land, flight instructor. Also offers bachelors degree in aviation.

Alaska

Anchorage Community College
Office of Admissions
2533 Providence Ave.
Anchorage, AK 99508
Program: Private, commercial, instrument, multiengine, flight instructor. Also offers two year associate degree in professional piloting.

Tanana Valley Community College
Office of Admissions
Fairbanks, AK 99701
Program: Private, commercial, instrument, multiengine, flight instructor, ATP, seaplane. Also offers two year associate degree in professional piloting.

Arizona

Arizona State University
Office of Admissions
Moeur, Room 136
Tempe, AZ 85281
Program: Private, commercial, instrument, multiengine, flight instructor.

Cochise College
Office of Admissions
Douglas, AZ 85607
Program: Private through ATP. Also offers two year associate degree in professional piloting.

Arkansas

Henderson State University
Office of Admissions
1100 Henderson St.
Arkadelphia, AR 71923
Program: Private through ATP. Also offers bachelors degree in aviation.

Mississippi County Community College
Office of Admissions
PO Box 1109
Blytheville, AR 72315
Program: Private, Commercial, Instrument. Associate degree in Aviation.

California

Chaffey College
Office of Admissions
5885 Haven Ave.
Alta Loma, CA 91701
Program: Private, commercial, instrument. Also offers two year associate degree in aviation.

College of San Mateo
Office of Admissions
1700 West Hillsdale Blvd.
San Mateo, CA 94402
Program: Private, commercial, instrument. Also offers two year associate degree in aviation as commercial pilot.

Cypress College
Office of Admissions
9200 Valley View
Cypress, CA 90630
Program: Private through flight instructor incorporated into college degree. Also offers one year certificate program for professional pilot and two year associate degree for commercial pilot.

Foothill College
Office of Admissions
12345 El Monte Rd.
Los Altos Hills, CA 94022
Program: Private through flight instructor. Also offers two year associate degree in aviation.

Glendale Community College
Office of Admissions
1500 N. Verdugo Rd.
Glendale, CA 91208
Program: Private, commercial, instrument, flight instructor. Also offers certificate program and two year associate degree in pilot training.

Grossmont College
Office of Admissions
8800 Grossmont College Dr.
El Cajon, CA 92020
Program: Private, commercial, instrument.

Long Beach City College
Office of Admissions
4901 E. Carson St.
Liberal Arts Campus
Long Beach, CA 90808
Program: Private, commercial, instrument. Also offers two year associate degree as professional pilot.

Merced College
Office of Admissions
3600 M. Street
Merced, CA 95340
Program: Private through flight instructor.

Miracosta Community College

Office of Admissions
One Barnard Drive
Oceanside, CA 92056
Program: Private, commercial, instrument. Also offers one year certificate program in pilot training.

Monterey Peninsula College

Office of Admissions
980 Fremont
Monterey, CA 93940
Program: Private, commercial, instrument. Also offers 1 year certificate and/or two year associate degree in aviation.

Mount San Antonio College

Office of Admissions
1499 N. State St.
Walnut, CA 91789
Program: Private-Flight instructor: airplane, instruments, multiengine-land. Also offers two year associate degree as commercial pilot.

Ohlone College

Office of Admissions
43600 Mission Blvd.
Fremont, CA 94539
Program: Private, commercial, instrument, flight instructor-airplane. Also offers one year certificate program and/or two year associate degree as commercial pilot.

Orange Coast College

Office of Admissions
2701 Fairview Rd.
Costa Mesa, CA 92626
Program: Private, commercial, instrument, flight instructor-airplane. Also offers two year associate degree in air transportation/commercial pilot.

San Bernardino Valley College

Office of Admissions
701 S. Mt. Vernon Ave.
San Bernardino, CA 92410
Program: Private, commercial, instrument.

San Diego Mesa College

Office of Admissions
7250 Mesa College Dr.
San Diego, CA 92111
Program: Private, commercial, instrument, multiengine land, flight instructor-airplane.

Sierra Academy of Aeronautics

Office of Admissions
Oakland International Airport
Oakland, CA 94614
Program: Private-ATP. Most ratings include helicopter. Program also includes Cessna Citation ATP/Type rating, Rockwell Sabreliner ATP, Boeing 707 Flight Engineer and ATP, Boeing 727 Flight Engineer and ATP, and Boeing 737 ATP.

Colorado

Aims Community College

Office of Admissions
PO Box 69
Greeley, CO 80632
Program: Private, commercial, instrument, multiengine land, flight instructor for airplane and instruments. Also offers two year associate degree in aviation technology.

Colorado Northwestern Community College

Office of Admissions
500 Kennedy Drive
Rangely, CO 81648
Program: Private through flight instructor airplane and instrument. Also offers two year associate degree in aviation occupations (career pilot program).

Metropolitan State College

Office of Admissions
1006 11th St.
Denver, CO 80204
Program: Private (airplane and helicopter), through ATP. Also offers bachelors degree in professional piloting.

United States Air Force Academy
Office of Admissions
USAF Academy, CO 80840
Program: Private through flight instructor—instruments.

Connecticut

University of New Haven
Office of Admissions
300 Orange Avenue
West Haven, CT 06516
Program: Private through flight instructor-airplane. Also offers two year associate degree in aeronautics (flight).

Delaware

Wilmington College
Office of Admissions
320 DuPont Hwy
New Castle, DE 19720
Program: Private through flight instructor-multiengine.

Florida

Broward Community College
Office of Admissions
225 East Las Olas Blvd.
Ft. Lauderdale, FL 33301
Program: Private through flight instructor—instruments. Also offers two year associate degree as career pilot.

Embry Riddle Aeronautical University
Office of Admissions
Star Route, Box 540
Bunnell, FL 32010
Program: Private through flight instructor—multiengine. Also offers jet type rating—Flight engineer B 727. You can also take a 1¹/₂ year program of intensified flight training towards your flight certificates.

Flight Safety International
Vero Beach Municipal Airport
PO Box 2708
Vero Beach, FL 32960
Program: Private through ATP.

Florida Institute of Technology
Office of Admissions
PO Box Drawer 1839
Melbourne, FL 32901
Program: Private through flight instructor—multiengine land.

Florida Junior College at Jacksonville
Office of Admissions (Kent Campus)
3939 Roosevelt Blvd.
Jacksonville, FL 32205
Program: Private, commercial, instrument.

Gulf Coast Community College
Office of Admissions
5230 West Highway 98
Panama City, FL 32401
Program: Private through ATP. Also offers two year associate degree as career pilot.

Manatee Jr. College
Office of Admissions
5840 26th St. West
Bradenton, FL 33506
Program: Private, commercial, instrument.

Miami-Dade Community College
Office of Admissions
South Campus
11011 SW 104 St.
Miami, FL 33167
Program: Private through flight instructor-multiengine. Also offers two year associate degree as career pilot.

Palm Beach Jr. College
Office of Admissions
4200 Congress Ave.
Lake Worth, FL 33461

Program: Private, commercial, instrument. Also offers two year associate degree as career pilot.

St. Petersburg Junior College
Office of Admissions
PO Box 13489
St. Petersburg, FL 33733
Program: Private through flight instructor—multiengine.

Hawaii

Honolulu Community College
Office of Admissions
874 Dillingham Blvd.
Honolulu, HI 96817
Program: Private, commercial, instrument.

Idaho

Aero Technicians, Inc.
Rexburg, ID 83440
Program: Private through ATP.

Illinois

Belleville Area College
Office of Admissions
2500 Carlyle Rd.
Belleville, IL 62221
Program: Private through flight instructor—multiengine. Program includes one year certificate program and two year associate degree.

Carl Sandburg College
Office of Admissions
2232 So. Lake Storey Rd.
Galesburg, IL 61401
Program: Private, commercial, instrument.

Moody Bible Institute
Office of Admissions—Coordinator
820 N. LaSalle St.
Chicago, IL 60610

Program: Private (Airplane and Helicopter) through ATP. Also offers glider, seaplane (private, commercial, ATP).

Parks College of St. Louis University
Office of Admissions
Administration Building
Cahokia, IL 62206
Program: Private through flight instructor—multiengine land. Also offers bachelors degree as professional pilot.

Rock Valley College
Office of Admissions
3301 N. Mulford Rd.
Rockford, IL 61111
Program: Private, commercial, instrument. Also offers two year associate degree in commercial aviation.

Southern Illinois University at Carbondale
Office of Admissions
Woody Hall
Carbondale, IL 62901
Program: Private through ATP. Also includes DC3 and Lear jet type ratings.

Thornton Community College
15800 South State Street
South Holland, IL 60473
Program: Private, Commercial, Instrument. Certificate, associate degree in aviation.

University of Illinois
Office of Admissions
108 Administration Building
Urbana, IL 61801
Program: Private (airplane and helicopter) through ATP. Also offers two year certificate program as a professional pilot.

Indiana

Indiana State University
Office of Admissions
217 N. 6th St.
Terre Haute, IN 47809

Program: Private through flight instructor—multiengine land. Also offers bachelors degree as professional pilot.

Purdue University
Office of Admissions
Hovde Hall
West Lafayette, IN 47907
Program: Private through ATP. Also Turbojet Flight Engineer, B 707.

Vincennes University
Office of Admissions
1002 First St.
Vincennes, IN 47591
Program: Private through flight instructor—multiengine land. Also Agricultural flight training.

Iowa

Iowa Western Community College
Office of Admissions
2700 College Rd. Box 4-C
Council Bluffs, IA 51502
Program: Private (airplane and helicopter) through flight instructor—multiengine.

University of Dubuque
Office of Admissions
2050 University
Dubuque, IA 52001
Program: Private through flight instructor—multiengine.

Kansas

Central College
1200 S. Main Street
McPherson, KS 67460
Program: Private, Commercial, Instrument, Flight Instructor. Also offers associate degree in aviation.

Garden City Community College
Office of Admissions
801 Campus Drive
Garden City, KS 67486

Program: Private through flight instructor—instruments. Also offers two year associate degree in flight training.

Hesston College
Office of Admissions
Administration Bldg. Box 3000
Hesston, KS 67062
Program: Private through flight instructor—multiengine. Also offers two year associate degree in aviation.

Neosho County Community College
1000 So. Allen
Chanute, KS 66720
Program: Private through ATP. Also offers two year associate degree in aeronautical science.

University of Kansas
Office of Admissions
Lawrence, KS 66045
Program: Private through flight instructor—multiengine.

Louisiana

Louisiana Tech University
Office of Admissions
PO Box 5226
Ruston, LA 71272
Program: Private through ATP. Also offers bachelors degree in professional aviation.

Nicholls State University
Office of Admissions
PO Box 2022
Thibodaux, LA 70310
Program: Private through flight instructor—multiengine. Also Seaplane. Also offers two year associate degree in aeronautical science.

Northeast Louisiana University
Office of Admissions
700 University Ave.
Monroe, LA 71209
Program: Private through ATP.

Northwestern State University
Office of Admissions
Natchitoches, LA 71457
Program: Private through ATP. Also offers bachelors degree in aviation science.

Maryland

Catonsville Community College
Office of Admissions
800 So. Rolling Rd.
Catonsville, MD 21228
Program: Private and commercial. Also offers two year associate degree in pilot training.

Massachusetts

Central New England College
Office of Admissions
768 Main Street
Worcester, MA 01610
Program: Private through flight instructor—multiengine-land.

North Shore Community College
Office of Admissions
3 Essex Street
Beverly, MA 01915
Program: Private, commercial, instrument. Also offers two year associate degree in aviation science technology.

Northeastern University/Lincoln College
Office of Admissions
South Bedford Rd.
Burlington, MA 01803
Program: Private through flight instructor—instrument. Also offers two year associate degree in aviation technology.

Michigan

Andrews University
Office of Admissions
Berrien Springs, MI 49104
Program: Private through flight instructor—multiengine-land.

Jackson Community College
Office of Admissions
2111 Emmons Rd.
Jackson, MI 49201
Program: Private through flight instructor—instruments. Also offers two year associate degree in aviation technology.

Lansing Community College
Office of Admissions
PO Box 40010
430 North Capitol Ave.
Lansing, MI 48901
Program: Private through flight instructor—multiengine-land.

Northern Michigan University
Office of Admissions
304 Cohodas Adm. Bldg.
Marquette, MI 49855
Program: Private through ATP. Private, commercial, instrument and flight instructor—airplane are incorporated into associate degree program. Also offers two year associate degree in aviation technology.

Northwestern Michigan College
Office of Admissions
1701 E. Front St.
Traverse City, MI 49684
Program: Private through flight instructor-multiengine land. Also offers two year associate degree in flight technology.

Western Michigan University
Office of Admissions
Administration Bldg.
Kalamazoo, MI 49008
Program: Private through flight instructor—instrument. Also offers bachelors degree in flight technology.

Minnesota

Mankato State University
Office of Admissions
Box 55
Mankato, MN 56001

Program: Private through flight instructor—airplane. Private, commercial and instrument ratings incorporated into University degree program.

St. Cloud State University
Office of Admissions
Admin. Services Rm. 118
St. Cloud, MN 56301
Program: Private through ATP. Also offers bachelors program in engineering technology with a flight option.

University of Minnesota Technical College
Office of Admissions
Selvig Hall
Crookston, MN 56716
Program: Private, commercial, instrument, agriculture-aerial applicator training.

University of Minnesota—Twin Cities
Office of Admissions
231 Pillsbury Drive SE
Minneapolis, MN 55455
Program: Private, commercial, instruments, multiengine-land, and flight instructor—airplane.

Vermillion Community College
Office of Admissions
1900 East Camp Street
Ely, MN 55731
Program: Private through Commercial-Instrument. Associate in professional piloting, wilderness pilot.

Winona State University
Office of Admissions
125 Phelps Hall
Winona, MN 55987
Program: Private, commercial, instrument.

Missouri

Central Missouri State University
Office of Admissions
Warrensburg, MO 64093

Program: Private through ATP plus private and commercial helicopter.

Meramac Community College
Office of Admissions
11333 Big Bend Blvd.
St. Louis, MO 63122
Program: Private through flight instructor-airplane. Also offers two year associate degree as professional pilot.

Missouri Western State College
Office of Admissions
Student Services Bldg.
St. Joseph, MO 64507
Program: Private through flight instructor-airplane. Also offers two year associate degree in pilot training.

School of the Ozarks
Office of Admissions
Point Lookout, MO 65726
Program: Private through ATP. Also offers bachelors degree in aviation science.

Nebraska

Chadron State College
Office of Admissions
Crites Hall
Chadron, NE 69337
Program: Private through flight instructor—airplane. Also offers two year associate degree in air pilot.

New Hampshire

Daniel Webster College
Office of Admissions
University Drive
Nashua, NH 03063
Program: Private through ATP.

Nathaniel Hawthorne College
Office of Admissions
Antrim, NH 03440

Program: Private through ATP. Also type rating: DC3. Also offers bachelors degree in aeronautical science (flight).

New Jersey

Mercer County Community College
Office of Admissions
PO Box B
Trenton, NJ 08690
Program: Private, commercial, instrument. Also offers two year associate degree in flight technology.

New York

Dowling College
Office of Admissions
Idle Hour Blvd.
Oakdale, NY 11769
Program: Private through flight instructor-airplane. Also offers bachelors degree in aeronautics with flight option.

State University of New York
Agricultural and Technical College at Farmingdale
Office of Admissions
Adm. Bldg. Melville Rd.
Farmingdale, NY 11735
Program: Private, commercial, instrument, multiengine-land. Also offers two year associate degree in flight operations.

North Carolina

Guilford Technical Institute
Office of Admissions
PO Box 309
Jamestown, NC 27282
Program: Private through ATP (arranged on demand).

Lenoir Community College
Office of Admissions
PO Box 188
Kinston, NC 28501

Program: Private, commercial, instrument. Also offers two year associate degree in aviation management with a career pilot option.

Piedmont Bible College
Office of Admissions
716 Franklin St.
Winston-Salem, NC 27101
Program: Private through flight instructor—instruments, plus seaplane. Also offers bachelors degree in aviation education with a flight option.

North Dakota

University of North Dakota
Office of Admissions
Grand Forks, ND 58202
Program: Private through ATP. Also offers seaplane, glider certification and helicopter certification for anyone who has commercial-instrument airplane ratings. Also offers two year associate degree as professional pilot. FE, DC-9 Type four year bachelors degree in aviation.

Ohio

Cuyahoga Community College
Office of Admissions
Eastern Campus
4250 Richmond Rd.
Warrensville, Twp., OH 44122
Program: Private through flight instructor—instrument. Also offers two year associate degree in aviation technology.

Kent State University
Office of Admissions
145 Rockwell Hall
Kent, OH 44242
Program: Private through flight instructor-instruments., Also offers bachelors degree in aerospace flight technology.

Miami University
Office of Admissions
East High Street
Oxford, OH 45056
Program: Private, commercial, instrument. Also offers bachelors degree in aeronautics with flight option.

The Ohio State University
Office of Admissions
Third Floor Lincoln Tower
1800 Cannon Dr.
Columbus, OH 43210
Program: Private through ATP.

Ohio University
Office of Admissions
Chubb Hall
Athens, OH 45701
Program: Private through ATP. Also offers two year associate degree in aviation technology.

Oklahoma

American Flyers Inc.
Office of Admissions
PO Box 3241
Ardmore, OK 73401
Program: Private through ATP plus flight engineer: B 727.

Bethany Nazarene College
Office of Admissions
6729 N.W. 39th Expwy.
Bethany, OK 73008
Program: Private through multiengine-land. Also offers bachelors degree in aviation-business with a flight option.

Northeastern Oklahoma State University
Office of Admissions
Tahlequah, OK 74464
Program: Private through flight instructor-instruments.

Oklahoma State University
Office of Admissions
Whitehurst Hall
Stillwater, OK 74078
Program: Private through ATP.

Southeastern Oklahoma State University
Office of Admissions
Station A, Box 4139
Kuranti, OK 74701
Program: Private through ATP plus flight engineer—basic and turbojet. Also offers bachelors degree in aviation flight.

Spartan School of Aeronautics
Office of Admissions
8820 E. Pine St.
Tulsa, OK 74151
Program: Private through ATP.

University of Oklahoma
Office of Admissions
1000 Asp Ave.
Buchanan Hall
Norman, OK 73019
Program: Private through flight instructor-multiengine land. Also offers bachelors degree in aviation education with flight option.

Oregon

Lane Community College
Office of Admissions
4000 E. 30th Ave.
Eugene, OR 97405
Program: Private through ATP. Also offers two year associate degree in flight technology.

Mt. Hood Community College
Office of Admissions
26000 SE Stark St.
Gresham, OR 97030
Program: Private through flight instructor-instruments. Also offers two year associate degree in flight technology.

Treasure Valley Community College
Office of Admissions
650 College Blvd. Weese Bldg.
Ontario, OR 97914
Program: Private through ATP. Also offers two
year associate degree as career pilot.

Pennsylvania

United Wesleyan College
Office of Admissions
1414 E. Cedar St.
Allentown, PA 18103
Program: Private through flight instructor-
instruments.

South Carolina

Florence-Darlington Technical College
Office of Admissions
PO Drawer F-8000
Florence, SC 29501
Program: Private through multiengine-land.
Also offers two year associate degree in avia-
tion technology.

South Dakota

Augustana College
Office of Admissions
29th and Summit Ave.
Sioux Falls, SD 57197
Program: Private through flight instructor-
instruments. Also offers two year associate
degree as professional pilot.

Tennessee

Chattanooga State Technical College
Office of Admissions
4501 Amnicola Highway
Chattanooga, TN 37406
Program: Private through flight instructor-multiengine land. Also offers two year associate
degree in aerospace technology.

Cleveland State Technical College
Office of Admissions
Cleveland, TN 37311
Program: Private through multiengine-land.

Hiwassee College
Office of Admissions
Madisonville, TN 37354
Program: Private through multiengine land.
Also offers two year associate degree in aero-
nautical technology.

Middle Tennessee State University
Office of Admissions
Cope Administration Bldg.
Murfreesboro, TN 37132
Program: Private through ATP. Also offers
bachelors degree in aerospace technology with
flight option.

Steed College
Office of Admissions
Box 3098 CRS
Johnson City, TN 37601
Program: Private through ATP. Also offers two
year associate degree in aeronautical science.

Tennessee Wesleyan
PO Box 40
Athens, TN 37303
Program: Private through ATP with bachelors
degree in aviation.

University of Tennessee Space Institute
Office of Graduate Admissions
UTSI A203
Tullahoma, TN 37388
Program: Private through ATP.

Volunteer State Community College
Office of Admissions
Nashville Pike
Gallatin, TN 37066
Program: Private and instrument.

Texas

Central Texas College
Office of Admissions
Hwy 190 West
Killeen, TX 76542
Program: Private through ATP offered in both airplane, helicopter and seaplane. Also offers two year associate degree as a career pilot.

Letourneau College
Office of Admissions
PO Box 7001
Longview, TX 75607
Program: Private through flight instructor-multiengine land.

Mountain View College
Office of Admissions
4849 W. Illinois Ave.
Dallas, TX 75211
Program: Private through flight instructor-multiengine land. Also offers two year associate degree in aviation technology.

Navarro College
Office of Admissions
Box 1170
Corsicana, TX 75110
Program: Private through ATP. Also offers two year associate degree as career pilot.

Southwest Texas Junior College
Office of Admissions
Garner Field Rd.
Uvalde, TX 78801
Program: Private through flight instructor-airplane. Also offers two year associate degree in aviation technology.

Texarkana Community College
Office of Admissions
2500 N. Robinson Rd.
Texarkana, TX 75501
Program: Private through flight instructor multiengine land. Also offers two year associate degree in flight technology.

Texas State Technical Institute
Office of Admissions
Waco, TX 76705
Program: Private through flight instructor multiengine land.

Utah

Westminster College
Office of Admissions
1800 S. 13th, East
Salt Lake City, UT 84105
Program: Private through ATP. Also offers bachelors degree as professional pilot.

Virginia

Liberty University
3765 Candlers Mountain Road
Lynchburg, VA 24506
Program: Private, Commercial Instrument. Associate degree in aviation.

Washington

Big Bend Community College
Office of Admissions
Andrews & 24th St. Bldg. 1700
Moses Lake, WA 98837
Program: Private through flight instructor multiengine land.

Central Washington University
Office of Admissions
Mitchell Hall
Ellensburg, WA 98926
Program: Private through flight instructor-multiengine land plus seaplane.

Green River Community College
Office of Admissions
12401 SE Washington
Auburn, WA 98002
Program: Private through ATP.

Yakima Valley College
Office of Admissions
PO Box 1647
Yakima, WA 98902
Program: Private through flight instructor-multiengine land. Also offers associate degree in flight technology.

West Virginia

Salem College
Office of Admissions
223 West Main St.
Salem, WV 26426

Program: Private through flight instructor-multiengine land. Also offers bachelors degree in career aviation.

Wisconsin

Gateway Technical Institute
Office of Admissions
3520 30th Ave.
Kenosha, WI 53140
Program: Private through ATP. Also offers two year associate degree in pilot training.

Index

probationary year, 81-89
recommendations and resumes, 54-55
resume preparation, 52-54
savings accounts, 88-89
training, 21-23
vision requirements, 18, 19, 33, 92
women as, 20-21
Northwest Aerospace Training Corporation (NATCO), 26, 27, 45
Northwest Airlines, 6, 7, 9-14, 16, 26, 27, 29, 32, 45, 83, 84, 86, 93

P

Pan Am, 16, 44, 45
physical exams, 4, 17-20, 33, 68, 79-80, 92
Piedmont Airlines, 16
pilot domiciles, 17
Piloting Careers, 33
private pilot, 18, 22, 23, 26, 41, 42
probationary year, 81-89
  airplane checkout flight, 85-86
  final simulator check and evaluation, 88
  flight simulator training, 84, 85
  initial operating experience (IOE), 86-88
  training school, 82-88
  written tests, 84
Professional and Administrative Career Examination Guide, 72
proficiency checks, 4
psychological tests, 71
publications, 15

R

Ransome Airlines, 44, 45
recommendations (resumes), 54-55
rental aircraft rates, 21
Republic Airlines, 6, 10, 18, 83, 84, 86
Reserves, Military, 49
resume preparation, 52-54
retirement programs, 17
Rothmeir, Steve, 10

S

salary (*see* earnings and benefits)
savings accounts, 88-89
scheduling, 17
seniority systems, 6-8, 83, 84
Sierra Academy, 24
single-engine approach, flight simulator, 77
Southwest Airlines, 16, 38
stalls, flight simulator, 77
Stanine Test, 71
steep turns, flight simulator, 76
student pilot, 18, 20, 42

T

takeoffs, flight simulator, 76
training, 2, 21-23
  ab initio, 26-29
  airline-sponsored, 26, 27, 28, 29
  British Airways, 28-29
  civilian vs. military, 23-25, 48-50
  cockpit procedure trainers (CPT), 27

colleges and university, aviation programs, 95-107
cost of, 22
financing for, 23
flight instructors, 22
flight schools, 22, 24
flight simulators, 26, 27, 84, 85
FlightSafety Airline Transition Program, 46-48
ground school, 22
line-oriented flight training (LOFT), 28, 86
NATCO, 26, 27, 45
private pilot, 41
private pilot to airlines, 26, 41
rental aircraft for, rates, 21
turns, steep, flight simulator, 76
TWA, 16
two-tier pay, 1, 5, 6, 15, 16, 92

U

U.S. Air, 16
unions, 15, 83
United Airlines, 26, 29
UPS, 38

V

vision requirements, 18, 19, 33, 92

W

wages (*see* earnings and benefits)
women as airline transport pilot (ATP), 20-21
written tests, 84

Other Bestsellers of Related Interest

ILLUSTRATED ENCYCLOPEDIA OF GENERAL AVIATION—2nd Edition—Paul Garrison

From aerobatic flying to zulu time, this is your source for up-to-date details and information. This comprehensively revised edition includes all the latest facts on general aviation regulations, aircraft mechanics, weather, emergency procedures, avionics, and more. Terms are defined in alphabetical order, and more than 400 charts and illustrations are provided. There are even listings of aviation organizations, publications, airframe and avionics manufacturers, and federal and state aviation departments. 496 pages, 429 illustrations. Book No. 3316, $24.95 paperback only

IMPROVE YOUR FLYING SKILLS: Tips from a Pro—Donald J. Clausing

Learn firsthand and professional attitudes, flying standards, and everyday procedures practiced by airline and corporate aircraft captains. Leading off with an overview of the basics of flying VFR and IFR, the author gives in-depth coverage of all the things that make up advanced, no-nonsense airmanship. Among the topics covered: flight planning, cruise control, types of approaches, weather flying, filing IFR, radio procedures, and "BILAHS" (Briefing, IFR, Log, Alternate, Hazardous weather). 224 pages, 42 illustrations. Book No. 3328, $14.95 paperback, $24.95 hardcover

UNCONVENTIONAL AIRCRAFT—2nd Edition —Peter M. Bowers

From one of America's foremost aviation photographers and historians comes what is probably the largest collection of information on unconventional aircraft ever assembled. Far-out flying machines that have appeared in the last 80 years receive their due honor, humor, and respect in this salute to them and their creators. For this revised and expanded edition, Bowers has added 77 additional aircraft, including the Rutan Voyager, Beechcraft Starship, the Bell/Boeing V-22 Osprey tilt-rotor VTOL, and the Zeppelin-Staaken R-IV bomber. 336 pages, 437 illustrations. Book No. 2450, $19.95 paperback, $28.95 hardcover

ABCs OF SAFE FLYING—2nd Edition —David Frazier

Attitude, basics, and communication are the ABCs David Frazier talks about in this revised second edition of a book that answers all the obvious questions, and reminds you of others that you might forget to ask. This edition includes additional advanced flight maneuvers, and a clear explanation of the Federal Airspace system. 192 pages, 69 illustrations. Book No. 2430, $12.95 paperback only

GENERAL AVIATION LAW—Jerry A. Eichenberger

Although the regulatory burden that is part of flying sometimes seems overwhelming, it need not take the pleasure out of your flight time. Eichenberger provides an up-to-date survey of many aviation regulations, and gives you a solid understanding of FAA procedures and functions, airman ratings and maintenance certificates, the implications of aircraft ownership, and more. This book will allow you to recognize legal problems before they result in FAA investigations and the potentially serious consequences. 240 pages. Book No. 3431, $16.95 paperback, $25.95 hardcover

THE ILLUSTRATED GUIDE TO AERODYNAMICS—Hubert "Skip" Smith

If you've always considered aerodynamic science a highly technical area best left to professional engineers and aircraft designers. . .this outstanding new sourcebook will change your mind! Smith introduces the principles of aerodynamics to everyone who wants to know how and why aircraft fly. . .but who doesn't want to delve into exotic theories or complicated mathematical relationships. 240 pages, 232 illustrations. Book No. 2390, $17.95 paperback only

THE PILOT'S RADIO COMMUNICATIONS
HANDBOOK—3rd Edition—Paul E. Illman
and Jay Pouzar

". . .should have a spot on your bookshelf. . ."
—Private Pilot
". . .time spent on this book is sure to make your flight smoother." **—Kitplanes**

An updated edition of a popular handbook. Contains information on FAA rule changes regarding Mode C transponders, single-class TCA operations, and student pilot TCA training requirements. Current issues relating to the entire spectrum of VFR radio communications are addressed. Now, you can use even the busiest airports with confidence and skill. 240 pages, 61 illustrations. Book No. 2445, $15.95 paperback only

AM/FAR 1990—TAB/AERO Staff

Unsurpassed by any other in the industry, this pilot resource features New FAR Part 91 Rules—Complete and Unabridged—with a summary of proposed versions for easy reference. Federal regulations that directly impact general aviation, new information, extensive coverage of Loran-C, and much much more. Don't fly without it. 608 pages, Illustrated. Book No. 24390, $11.95 paperback, $19.95 hardcover

MORE I LEARNED ABOUT FLYING FROM
THAT—Editor of *FLYING*® magazine

What would you do if you lost half your wing in a midair collision or a strike force of wasps attacked you in the cockpit at 3,000 feet? What pilots have actually done in these nightmarish situations and in dozens of others is told by the fliers themselves in this gathering of accounts from *Flying* magazine's popular "I Learned About Flying From That" column. This is a treasury of unforgettable true tales and incisive pointers that save you from learning about flying the hard way. 196 pages. Book No. 3317, $12.95 paperback only

FLYING VFR IN MARGINAL WEATHER
—2nd Edition—Paul Garrison
Revised by Norval Kennedy

In this revised edition, you'll find technological information on such weather phenomena as wind shear . . .details on today's most advanced lightplane instrumentation. . .tips on the use of wing levelers and autopilots. . .and a practical look at the most advanced new technology in VHF navigation receivers, OBIs, and other navigation equipment including Loran C. 224 pages, 91 illustrations. Book No. 2416, $14.95 paperback only

GOOD TAKEOFFS AND GOOD LANDINGS
—Joe Christy

This reference thoroughly examines single-engine aircraft takeoffs and landings, and the critical transitions accompanying each. The author stresses that every pilot must continually evaluate ever-changing factors of wind, air pressure, precipitation, traffic, temperature, visibility, runway length, and braking conditions. *Good Takeoffs and Good Landings* belongs on every pilot's required reading list. 192 pages, 70 illustrations. Book No. 2487, $14.95 paperback only

YOUR PILOT'S LICENSE—4th Edition—Joe Christy

Completely revised, this book offers all the information on student training requirements, flight procedures, and air regulations. It tells you what the physical qualifications are, frankly discusses the expense involved, explains the integral role ground study plays in learning to fly, and even supplies a sample written test comparable to the actual Private Pilot's Written Test. Active pilots and flight instructors will find this an excellent refresher reference, too! 176 pages, 73 illustrations. Book No. 2477, $12.95 paperback only

STEALTH TECHNOLOGY: The Art of Black Magic
—J. Jones, Edited by Matt Thurber

Revealed in this book are the findings of careful research conducted by J. Jones, an expert in the area of military aviation stealth applications. The profound effect modern stealth technology is having throughout the United States and the world is discussed. Facts about the B-2 Stealth Bomber and the F117A Fighter are included. 160 pages, 87 illustrations, 4 pages full-color photographs. Book No. 22381, $14.95 paperback only

MAYDAY, MAYDAY, MAYDAY! Spin Instructions
Please!—Bob Stevens

Pilots, aviations buffs, and anyone with a sense of humor will enjoy this all-new collection of cartoons created by the undisputed Dean of American Aviation Cartooning, Bob Stevens. Poking fun at every aspect of general aviation, from soloing to flying the airways (and battling the bureaucracy) to the inevitable hangar flying, *Mayday* doesn't disappoint. 224 pages, Illustrated. Book No. 28964, $12.95 paperback only

Look for These and Other TAB Books at Your Local Bookstore

To Order Call Toll Free 1-800-822-8158
(in PA, AK, and Canada call 717-794-2191)

or write to TAB BOOKS, Blue Ridge Summit, PA 17294-0840.

Title	Product No.	Quantity	Price

☐ Check or money order made payable to TAB BOOKS

Charge my ☐ VISA ☐ MasterCard ☐ American Express

Acct. No. _____ Exp. _____

Signature: _____

Name: _____

Address: _____

City: _____

State: _____ Zip: _____

Subtotal $ _____

Postage and Handling
($3.00 in U.S., $5.00 outside U.S.) $ _____

Add applicable state
and local sales tax $ _____

TOTAL $ _____

TAB BOOKS catalog free with purchase; otherwise send $1.00 in check or money order and receive $1.00 credit on your next purchase.

Orders outside U.S. must pay with international money order in U.S. dollars.

TAB Guarantee: If for any reason you are not satisfied with the book(s) you order, simply return it (them) within 15 days and receive a full refund.

BC